T0197906

Nuclear Power: A Very Short Introduction

VERY SHORT INTRODUCTIONS are for anyone wanting a stimulating and accessible way into a new subject. They are written by experts, and have been translated into more than 45 different languages.

The series began in 1995, and now covers a wide variety of topics in every discipline. The VSI library now contains over 500 volumes—a Very Short Introduction to everything from Psychology and Philosophy of Science to American History and Relativity—and continues to grow in every subject area.

Titles in the series include the following:

Maxwell Irvine

NUCLEAR POWER
A Very Short Introduction

OXFORD
UNIVERSITY PRESS

OXFORD
UNIVERSITY PRESS

Great Clarendon Street, Oxford ox2 6DP
Oxford University Press is a department of the University of Oxford.
It furthers the University's objective of excellence in research,
scholarship, and education by publishing worldwide.

Oxford is a registered trade mark of Oxford University Press
in the UK and in certain other countries

First edition published 2011
Reprinted with corrections 2014

Published in the United States of America by Oxford University Press
198 Madison Avenue, New York, NY 10016, United States of America

British Library Cataloguing in Publication Data
Data available

Library of Congress Cataloging in Publication Data
Data available

Printed and bound by
CPI Group (UK) Ltd, Croydon, CR0 4YY

ISBN 978-0-19-958497-0

Preface

Energy is the single most important commodity for our survival and continuation of our society. With energy we can produce water and food. Energy is essential for modern agriculture and medicine. It is essential for everything we manufacture, for the buildings that we erect and the heat, light, and power that we use. We need energy for every form of travel and all our global communications. Since the dawn of history, we have generated the energy we need by burning hydrocarbons. Originally, it was vegetable matter together with vegetable oils and animal fats. The Industrial Revolution saw these traditional sources of energy become inadequate and we turned to the fossil fuels of coal, oil, and gas. Today, over 80% of all the energy that we consume still comes from the burning of hydrocarbons.

The burning of hydrocarbons produces emissions of carbon-bearing gases. The overloading of our atmosphere with these emissions is held responsible for global warming leading to global climate change. Climate change is not a new phenomenon on Earth; there have been several cycles of natural change, and these have had a devastating impact on life forms in the past, including the extinction of many species of plants and animals. Many scientists believe that the current global warming has been initiated by human activity and in particular by our production and use of hydrocarbon fuels. While there are still some people who feel that the current global warming is part of a natural cycle and not necessarily the product of human activity, it would seem prudent, whatever the case may be, to reduce our emissions of carbon gasses as rapidly as possible in order not to aggravate the situation further, and there is a strong case for a coherent international effort to achieve this.

There are other reasons why the search is on for alternative energy supplies. There are finite reserves of fossil fuels and many analysts are predicting that we are at, or close to, the point of peak production. Already the world's largest consumer of energy, the USA, which built its industrial might on the back of enormously rich resources of fossil fuels, is importing nearly two-thirds of the oil that it consumes. The birthplace of the Industrial Revolution, the UK, is facing the exhaustion of its North Sea oil and gas reserves a mere 40 years after the fields were first opened. Finally, the strategic importance of energy inevitably leads to geopolitical concerns. It could be argued that the rise in militant Islam can be traced back to the Anglo-Iranian Oil Company's refusal to negotiate with Iran's first democratically elected government over control of that country's oil supplies and the subsequent overthrow of that regime and the restoration of the Shah. Russia, as the largest supplier of gas to Eastern Europe, has already created concerns in some of its former USSR client states over gas supplies, and such concerns in Finland led it to seek alternative energy supplies. The USA is facing pressure from its Latin and South American oil suppliers seeking a redress of the balance of power in that region.

At the close of the Second World War, the world became aware of the awesome power of a new source of energy – nuclear power. The initial euphoria engendered by the discovery of this new energy source quickly gave way to a cold realization that the technology had a long way to develop. Public concerns grew. Initially, these concerns originated from the fear of explosive accidents coupled to a general ignorance of a mysterious new phenomenon, radioactivity, and its links to cancer. Additional geopolitical concerns concentrated on the link to nuclear weapons proliferation. These concerns began when first India and then Pakistan demonstrated their nuclear weapons capacity and have not decreased since North Korea tested a nuclear weapon device. News that Iran is building a uranium-enrichment plant with the possibility that it might develop its own weapons programme greatly raises the stakes in the tense relationship between Israel (already widely believed to have a nuclear weapon capacity) and the Arab world.

The public were kept in the dark with regard to the first generation of 'civil' nuclear power stations which, while they did produce electricity for civil consumption, had their efficiency compromised by their covert objective of producing plutonium for Cold War weapons. The cost of nuclear power rose as improved safety requirements were imposed on the developing technology. In the UK and USA, every new reactor was a prototype, and there were no economies of scale from mass production. Each variation in design required separate extensive and expensive licensing arrangements. It took the French to demonstrate that nuclear power could be made economically competitive with fossil fuels through mass production and model safety licensing similar to that used for aircraft.

The emerging industry was hit by two events that seemed to confirm the worst of the public's fears. First, in 1979 a newly commissioned reactor at Three Mile Island, Pennsylvania, went out of control, and for a week the world waited for a catastrophic release of deadly radioactivity which never occurred. In 1986, the catastrophe that everyone feared did occur at Chernobyl in the Ukraine. These incidents led many countries to abandon their plans to develop nuclear power.

The public fears of nuclear power have frequently been fuelled by a less than sober press which seldom has followed up its alarmist headlines with the mundane explanations of the incidents that they have reported.

As the world grows aware of the approaching energy crunch, many countries are returning to the nuclear option, and with it concerns about decommissioning costs, radioactive waste management, and reactor safety are resurfacing, and have been refuelled by the impact of the 2011 earthquake- and tsunami-induced nuclear incident in Japan. It is not my purpose to deny that nuclear power poses risks; all human activities carry risk. I hope to put these risks into perspective and to provide the reader with information to better judge their impact and to offset the more alarmist media reporting.

This *Very Short Introduction* to nuclear power traces the evolution of nuclear science during the first half of the 20th century. It tells the story of the explosive development of nuclear technology that took place during the Second World War and resulted in the capacity to begin the evolution of civil nuclear reactors in the post-war years up to the present time. We discuss the development of the next generation of reactors. We examine the concerns for nuclear safety and place it in perspective in relation to other human activities. We present the lessons to be learned from Three Mile Island and Chernobyl, and discuss the issues around decommissioning and waste disposal. The cost of nuclear power is examined and comparisons with other energy sources presented. We take a brief glimpse at a future when thermal fission power may be replaced by nuclear fusion. Finally, we conclude with a discussion of energy issues facing the globe and analyse the need for a nuclear component in their solution.

I am grateful to my colleagues on the Royal Society of Edinburgh's Inquiry into Energy Issues in Scotland, the European Academies Science Advisory Committee's Energy Group, and the UK's Committee on Radioactive Waste Management Committee for sharing their insights with me. I am indebted to Sir Christopher Llewellyn Smith, Director of the JET Laboratory, for keeping me up to date on the development of fusion power, and to Neil Kermode and his staff at the European Marine Energy Centre for the opportunity to learn about the prospects for wave and tidal power.

The publisher and author would like to thank Professor Eric Finch for his assistance with the corrections to this title.

Above all, I thank my friends and family, who have put up with my seeming inability to discuss anything but energy issues for the past ten years.

Maxwell Irvine,
Manchester, 2011

Contents

Acknowledgements

There are a number of national government agencies and international intergovernmental agencies that gather data and prepare analytical reports about energy issues on a regular basis. Some specifically concentrate on nuclear issues. Together, these organizations provide a wealth of material on which I have depended in preparing this *Very Short Introduction* to nuclear power. Principal amongst these bodies are:

The International Atomic Energy Agency (IAEA) established in 1957 under the umbrella of the United Nations Organization to promote international cooperation in the peaceful development of nuclear technology. They are a lead organization in advising on nuclear safety and the principal monitors of adherence to the Nuclear Weapons Non-Proliferation Treaty.

The International Energy Agency (IEA) was established by the Organisation for Economic Co-operation and Development (OECD) in the aftermath of the 1974 oil crisis. The IEA is a principal source of data on energy sources, production, demand, and consumption.

The Nuclear Energy Agency (NEA) was also established under the umbrella of the OECD, originally as the European Nuclear Energy Agency in 1954, and changed its name to NEA in 1972 with the admission of non-European states. The mission of the NEA is to 'assist its members in maintaining and further developing through international cooperation the scientific, technological and legal basis for the safe, environmentally friendly and economical use of nuclear energy for peaceful purposes'. The NEA member states between them represent over 80% of globally installed nuclear capacity.

The Energy Information Administration (EIA) is part of the United States Department of Energy (DOE). In addition to statistics on current supply and demand for energy in the USA, the EIA prepares annual forward projections for the USA and the rest of the world.

Reports from these organizations and others feature highly in the Further Reading list.

List of illustrations

Illustrations have been drawn by the author unless otherwise credited.

List of tables

Chapter 1
A new science is born

By the end of the 19th century, it was clear that matter had an
atomic structure and that relative weight could be attached to the
atoms of different chemical elements. Many of these atomic
weights were close to being integer multiples of the weight of the
lightest element, hydrogen. However, there were many significant
exceptions to this. It was known that the atoms contained electric
charges and that, since most matter is electrically neutral, there
must be a balance of positive and negative charges. The precise
structure of these atoms and the reasons for their exact masses was
a complete mystery. Chemists had developed a table listing the
elements labelled by their measured atomic masses and electronic
charges (Figure 1). This demonstrated a periodicity in the chemical
properties of elements.

In general, only the naturally occurring elements are displayed in
Figure 1. With the development of nuclear reactors and
accelerators more than 20 man-made elements have been added to
the actinide series – these are the transuranic elements. The
elements are designated by their electronic charge numbers.

At the end of the 19th century, a time of tremendous scientific
advance, two developments were to transform the situation. First,
the discovery that certain materials emitted radiation, a
phenomenon now called 'radioactivity', was to provide scientists

1 H																		2 He
3 Li	4 Be											5 B	6 C	7 N	8 O	9 F	10 Ne
11 Na	12 Mg											13 Al	14 Si	15 P	16 S	17 Cl	18 Ar
19 K	20 Ca	21 Sc	22 Ti	23 V	24 Cr	25 Mn	26 Fe	27 Co	28 Ni	29 Cu	30 Zn	31 Ga	32 Ge	33 As	34 Se	35 Br	36 Kr
37 Rb	38 Sr	39 Y	40 Zr	41 Nb	42 Mo	43 Tc	44 Ru	45 Rh	46 Pd	47 Ag	48 Cd	49 In	50 Sn	51 Sb	52 Te	53 I	54 Xe
55 Cs	56 Ba	57–70 *	71 Lu	72 Hf	73 Ta	74 W	75 Re	76 Os	77 Ir	78 Pt	79 Au	80 Hg	81 Tl	82 Pb	83 Bi	84 Po	85 At	86 Rn
87 Fr	88 Ra	89– **															

* Lanthanide series

57 La	58 Ce	59 Pr	60 Nd	61 Pm	62 Sm	63 Eu	64 Gd	65 Tb	66 Dy	67 Ho	68 Er	69 Tm	70 Yb

** Actinide series

89 Ac	90 Th	91 Pa	92 U

1. The periodic table of chemical elements

with probes to study subatomic structures. Second, the development of Einstein's theory of relativity and, shortly after, the emergence of quantum theory were to provide the intellectual framework for interpreting the structures that were revealed.

Einstein's theory of relativity revealed that 'mass' was simply another manifestation of 'energy'. His equation $E = Mc^2$, where E is the energy equivalent to a mass M and c is the velocity of light, became one of the most famous formulas in science.

There were three types of radioactive radiation discovered. The first to be identified were beams of exceptionally light, negatively charged particles, now known as electrons, discovered by the British physicist J. J. Thomson in 1897.

Thomson first identified electrons in experiments in a Crookes tube, a precursor to the modern TV cathode ray tube. Further study revealed that the electrons seen by Thomson were identical to the light, negatively charged particles emitted in radioactive decay.

In a series of studies begun in 1898, the New Zealand scientist Ernest Rutherford, working with Frederick Soddy at McGill University in Canada, separated the radiation from radium into components according to their ability to penetrate matter and cause ionization. The least penetrative radiation, which Rutherford named 'alpha rays', had a positive electric charge. The electrons he called 'beta rays'. Because alpha and beta rays had electric charges, their trajectories could be manipulated by passing them through electromagnetic fields. By studying these tracks, Rutherford was able to deduce relative charges and masses for the radiation. The electron was the lightest object to have ever been identified. Its mass was only 0.511MeV, or approximately 1/2000th of the atomic mass of hydrogen. (In the subatomic world, it is usual to quote masses in energy units using Einstein's formula.) The energy standard is the electron-volt (eV), and its electric charge was -1.6×10^{-19} Coulombs. The alpha rays had a mass similar to that of the helium atom at 3,784MeV, approximately four times heavier than hydrogen. The alpha particles had a positive charge that was twice the magnitude of that of the electron.

In 1900, the French scientist Paul Villard discovered an electrically neutral form of radiation similar to the X-rays discovered by Röntgen in 1895 which revealed them to be a particularly high-frequency form of electromagnetic radiation. Using Planck's energy frequency relationship $E = h\nu$, where h is Planck's constant and ν is frequency, X-rays corresponded to energies in the range 1 to 100keV, while Villard's radiation was in the MeV range; Rutherford named these as 'gamma rays' in 1903.

The least penetrative radiation, the alpha rays, could be stopped in a sheet of cardboard; the beta rays were halted by a thin sheet of aluminium; while the most penetrative radiation, the gamma rays, required a high-density material like lead to halt their flow.

The alpha and beta particles carried an electric charge and hence could be accelerated in electric fields to produce more penetrative forms of radiation.

In 1907, Rutherford moved to the University of Manchester and continued his probing of matter with beams of alpha radiation. Working with an extraordinary group of exceptional scientists, Rutherford laid the foundations of the new science of nuclear physics.

First, in collaboration with Hans Geiger, he developed a detector for individual alpha particles. Then with John Nuttall and Ernest Marsden, he studied the scattering of alpha particle beams off thin gold films. In these experiments, some of the alpha particles were reflected back from the gold film. This indicated that they had encountered an object much more massive than themselves that had repelled them. The only things in the gold film were atoms of gold with an atomic mass number 190, almost 50 times heavier than the alpha particles. The atom was known to contain electric

2. Ernest Rutherford. Known as the 'father of nuclear science', Ernest Rutherford was born in Brightwater, New Zealand, 1871. He was awarded the Nobel Prize in 1908, while at the University of Manchester, for his investigations into the disintegration of the elements, and the chemistry of radioactivity. He died in Cambridge, UK, in 1937

charges, and Rutherford assumed that the repulsion was due to the familiar interaction between like charges, the positively charged alpha particles had encountered the source of the positive charge in the gold atom and it was associated with most of the mass of the atom.

Rutherford then developed a formula for the distribution of scattered alpha particles by a point concentration of positive charge which accurately reproduced the observations and allowed an estimate of the mass of the scattering centre. Thus Rutherford had demonstrated that all the positive charge in the gold atom was concentrated in a central core containing virtually all the mass of the atom.

This implied the existence of a new force in nature to add to the familiar classical forces of gravitation and electromagnetism. This force had to be attractive and much stronger than the electric repulsion between like charges in order to hold the nucleus together. It also had to be of extremely short range so as not to interfere in the scattering process.

If this was indeed the case, then Rutherford had a tool for obtaining an approximate measure of the size of this concentrated charge. The repulsive force between like electric charges is proportional to the two charges and inversely proportional to the square of the distance between them. However, at very short distances the force must change to reflect the stronger attractive force holding the positive charges in the atom together. Thus the Rutherford scattering formula should break down for alpha particle energies above E_0 (Figure 3), the energy at which the distance of closest approach was equal to the radius of the concentration of the positive charges, R_A. This was exactly what was observed. The early experiments by Rutherford and his colleagues in Manchester were restricted to alpha particles with the natural decay energy from radioactive elements. This meant that they could not explore the size of the nuclei but simply place an upper limit on their extent. When Rutherford moved to Cambridge, he was joined by John Cockcroft and Ernest Walton.

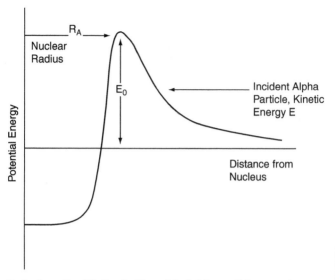

3. A schematic of Rutherford's model of alpha particle scattering

In the early 1930s, Cockcroft and Walton were responsible for the first major development and use of an accelerator for producing high energy charged particles. With the more flexible beams from accelerators, it became possible for scientists to obtain a clearer picture of nuclear sizes. The radii of nuclei were of order 10^{-15}m and increased with the cube root of the mass number, $R_A = r_0 A^{1/3}$, where r_0 was measured to be of order 10^{-15}m. This is consistent with the close packing of a number of particles each of the same size r_0.

The curve in Figure 3 represents the potential energy of interaction between the alpha particle and the nucleus. At large distances, this is the familiar repulsion between like charges. At short distances, less than R_A, the alpha particle experiences the nuclear attractive force. Rutherford developed a classical, that is, pre-quantum theory, formula for the scattering of alpha particles from a point nucleus. For incident alpha particle energies E less than E_0, the formula accurately replicated his experimental observations. For

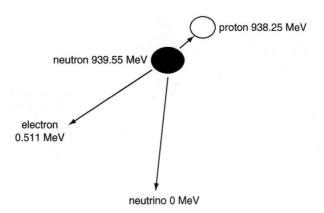

4. Neutron beta decay, showing particle masses

alpha particles (charge +2e) scattering from gold (charge 79e and mass number A=197), E_0 = 5.5MeV and R_A =5.8x10^{-15}m approximately.

It is now known that the atomic nucleus can be identified by two numbers: a mass number (A) and an atomic number (Z). For hydrogen A=Z=1. The hydrogen nucleus is called the 'proton'. For almost all heavier, stable nuclei, A is equal to or greater than Z. This is because except for A=Z=1 the nucleus also contains electrically neutral particles, called neutrons, similar in mass to the protons. Assuming the number of neutrons is N, then A = N + Z. It was James Chadwick, working with Rutherford in Cambridge, who in 1932 finally identified the neutron. Its mass is 939.55 MeV compared with the proton mass at 938.256 MeV. The collective name for protons and neutrons is nucleons.

The discovery of the neutron answered two problems. First, atoms of the same Z (chemical element) could exist with different numbers of neutrons. These are called 'isotopes', meaning at the same place in the chemical periodic table. Thus a chemical sample could include a mixture of isotopes and the chemically measured atomic weight would then be the weighted average of the nuclear mass numbers and hence not an integer. For example, chlorine has $Z = 17$ and the isotopes chlorine-35, -36, and -37 occur in nature with abundances 75.77%, traces, and 24.23%, respectively, leading to the atomic mass of 35.45. Second, since the neutron mass was greater than the proton mass by 1.29MeV, which is more than the mass of the electron, it was possible to identify the source of beta decay as the decomposition of a neutron into a proton plus an electron. However, if this was the case then the electrons (beta particles) should all emerge with an energy of 0.78MeV. In fact, the electrons appeared with a range of energies all less than 0.78MeV, suggesting that some as yet unidentified object was simultaneously emitted.

This object had to be electrically neutral and of almost zero mass. In 1930, Wolfgang Pauli had postulated the existence of such a particle in order to explain the spectrum of beta decay energies. With Chadwick's confirmation of the neutron hypothesis, Enrico Fermi named this new particle the 'neutrino', or little neutron (Figure 4).

In 1930, Paul Dirac had speculated that the relativistic version of quantum mechanics seemed to allow for the existence of antimatter, that is, particles identical to those familiar in the laboratory but with their properties reversed. In 1932, Carl Anderson found cosmic ray tracks that looked like an electron but with a positive electric charge. This was the first antiparticle and was named the positron. While it is energetically impossible for a free proton to emit a positron and change itself into a neutron, within the nucleus an isotope could be formed with a depletion of neutrons such that this form of beta decay could be realized. These two forms of beta decay are designated β^- and β^+ depending on whether an electron or positron is emitted.

It soon became clear that beta decay that resulted in the emission of an electron was accompanied by the emission of an antineutrino, while a positron was accompanied by the emission of a neutrino.

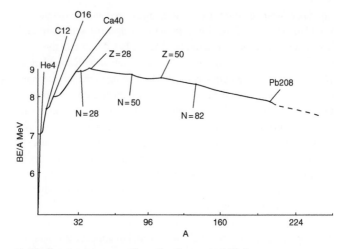

5. **Binding energy per nucleon for the most stable isotopes as a function of mass number A. The broken line indicates that all nuclei heavier than lead-208 are unstable**

With virtually no mass and no electric charge, the detection of neutrinos is extremely difficult. Neutrinos are the most penetrative radiation yet detected, and antineutrinos are the dominant form of radiation emitted by a nuclear reactor. In 1956, a US group led by Cowen and Reines was able to detect the antineutrinos from a reactor for the first time.

According to Einstein's mass–energy equation, the mass of any composite stable object has to be less than the sum of the masses of the parts; the difference is the binding energy of the object. In Figure 5, we display the measured binding energy per nucleon for the most stable isotopes. We see a generally smooth curve rising sharply in the light nuclei reaching a maximum of 8.8MeV for iron-56 before falling to a value of 7.6MeV for the heaviest naturally occurring isotope of uranium-238. All nuclei above lead-208 are radioactively unstable.

Among the very light nuclei, there are sharp peaks indicating significantly greater stability for the alpha particle (helium-4), the isotopes carbon-12 and oxygen-16. Other particularly stable nucleon numbers are indicated.

The general features of the binding energies are simply understood as follows.

We have seen that the measured radii of nuclei increased with the cube root of the mass number A. This is consistent with a structure of close packed nucleons. If each nucleon could only interact with its closest neighbours, the total binding energy would then itself be proportional to the number of nucleons. However, this would be an overestimate because nucleons at the surface of the nucleus would not have a complete set of nearest neighbours with which to interact (Figure 6). The binding energy would be reduced by the number of surface nucleons and this would be proportional to the surface area, itself proportional to $A^{2/3}$. So far we have considered only the attractive short-range nuclear binding. However, the protons carry an electric charge and hence experience an electrical repulsion between each other. The electrical force between two protons is much weaker than the nuclear force at short distances but dominates at larger distances. Furthermore, the total electrical contribution increases with the number of pairs of protons.

The main characteristics of the empirical binding energy of nuclei exhibited in Figure 5 can now be explained. For the very light nuclei, all the nucleons are in the surface, the electrical repulsion is negligible, and the binding energy per nucleon, BE/A increases as the volume and number of nucleons increases. Next, the surface effects start to slow the rate of growth of the BE/A yielding a region of most stable nuclei near charge number Z = 28 (iron). Finally, the electrical repulsion steadily increases until we reach the most massive stable nucleus (lead-208). Between iron and lead, not only does the BE/A decrease so also do the proton to neutron ratios since the neutrons do not experience the electrical repulsion.

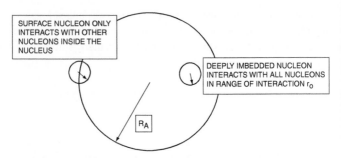

6. Nucleus of radius R_A containing nucleons with a range of interaction r_0. Deeply imbedded nucleons can interact with all other nucleons within the interaction radius. Nucleons at the surface can only interact with those within the nucleus. The number of surface nucleons is proportional to the surface area of the nucleus, $4\pi R_A^2$

This leaves the sharp peaks amongst the lightest nuclei and the residual peaks at specific numbers of neutrons and protons culminating in lead-208 ($Z = 82$, $N = 126$). For an explanation of these peaks, we must turn to the quantum nature of the problem. In explaining the electronic structure of the atom, Bohr had demonstrated that, while in classical physics there was a continuum of electronic orbits, in the quantum analysis only a restricted number of orbits could be realized. These orbits form shells with each shell having a specific number of orbits in it. Filled shells corresponded to particularly stable electronic structures identified as the inert rare gasses. This shell structure explained the periodicity of the chemical properties of elements. In the nuclear case, a shell structure also exists separately for both the neutrons and the protons. Closed-shell nuclei are referred to as 'magic number' nuclei. For example, the light nuclei helium-4 and oxygen-16 represent the first two doubly closed-shell nuclei. Other doubly closed-shell nuclei include calcium-40, calcium-48 and lead-208. While for the light nuclei helium-4 to calcium-40, these closed shells are occupied by equal numbers of neutrons and protons, as the nuclei get heavier the Coulomb repulsion term requires an increasing number of neutrons for stability, thus calcium-48 has $Z = 20$ and $N = 28$, while the heaviest stable nucleus, lead-208, has $Z = 82$ and $N = 126$. In the case of nuclei, closed-shell nuclei are once more particularly stable. The nuclear periodicity does not strictly follow the chemical periodicity but is once more explained by the quantum analysis of allowed nucleon orbits inside the nucleus.

Finally, if we look at the two nucleon systems, we find that the dineutron and the diproton (helium-2) are not bound, while the neutron–proton system called the deuteron, or heavy hydrogen (hydrogen-2), is bound and found in nature. Thus there is a particular stability for nuclei with equal numbers of protons and neutrons. This is particularly obvious for the light nuclei, and for the heavier nuclei it becomes a balancing act between this nuclear symmetry between neutrons and protons and the additional electrical repulsion between the protons, as illustrated in Figure 7.

Figure 7 displays the isotopes of nuclei. The 'most stable line' of black squares represents the isotopes of Figure 5. All nuclei above calcium-40 ($N = 20$, $Z = 20$) have a neutron excess in the most stable isotopes.

As we move off the line of stable nuclei, by adding or subtracting neutrons, the isotopes become increasingly less stable indicated by increasing levels of beta radioactivity. Nuclei with a surfeit of neutrons emit an electron, hence converting one of the neutrons into a proton, while isotopes with a neutron deficiency can emit a positron with the conversion of a proton into a neutron. For a heavy nucleus above lead-208, the neutron to proton ratio is reduced by a chain of beta decays by electron emission and alpha decays (see below). All nuclei heavier than lead-208 are unstable and hence radioactive alpha or β^- emitters.

Radioactive decay is a spontaneous event and when it will occur for a given isotope cannot be predicted. However, for a large sample of identical isotopes a mean lifetime can be defined. We shall consistently refer to the half-life, denoted by $t_{1/2}$. This is the

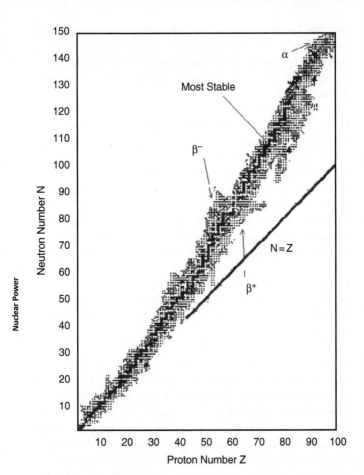

7. A chart of the nuclei

time over which half of the members of the sample will have
decayed. Thus after two cycles, only one-quarter of the original
isotopes will remain, and after three cycles only one-eighth will be
present, and so on. Lifetimes vary considerably; a free neutron
has a half-life of 15 minutes, tritium (hydrogen-3) 12.32 years,

14

radio-carbon (carbon-14) 6,000 years, and uranium-238 4.5 billion years.

The decay of uranium-238 is illustrative:

U^{238} -α-> Th^{234}-β^--> Pa^{234}-β^--> U^{234}-α-> Th^{230}-α-> Ra^{226}-α-> Rn^{222}-α-> Po^{218}-α-> Pb^{214}-β^--> Bi^{214}-β^--> Po^{214}-α-> Pb^{210}-β^--> Bi^{210}-β^--> Po^{210}-α-> Pb^{206} (stable)

Here the nomenclature indicate that uranium-238 decays by alpha particle emission to form thorium-234, which in turn emits an electron to become protactinium-234, and so on.

The fact that almost all the radioactive isotopes heavier than lead follow this kind of decay chain and end up as stable isotopes of lead explains this element's anomalously high natural abundance. The existence of isotopes heavier than lead in, for example, the uranium-238 decay chain is due to the long half-life of uranium-238, comparable to the age of the solar system.

We have seen that Rutherford was able to develop a classical scattering formula to explain alpha particle scattering from heavy nuclei. In 1928, George Gamow used the new quantum mechanics to describe alpha decay. The only possible source of the alpha particles was the nucleus. Quantum mechanics introduces the concept of the Heisenberg uncertainty principle which limits the precision with which the position and the momentum of a particle can simultaneously be known. If the alpha particle was constrained to be in the nucleus, its position was known to within the size of the nucleus (Figure 8). Quantum theory required the momentum of the alpha particle to be correspondingly uncertain. Thus while a classical particle would be trapped in the nucleus forever, a quantum particle would have an uncertain momentum and hence the possibility of escaping. Gamow calculated the relationship between the energy of an emitted alpha particle and

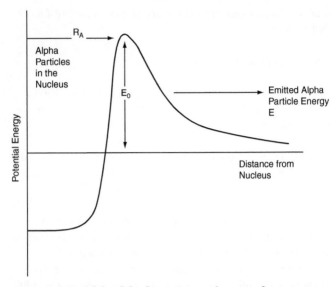

8. **Gamow's model for alpha decay. For uranium-238, the parameters are $R_A = 6.2 \times 10^{-15}$m and $E_0 = 5.9$MeV**

the half-life of the decay and was able to replicate the experimental results of Geiger and Nuttall.

Figure 5 clearly shows that as heavy nuclei decay, they release energy, and that if the light isotopes of hydrogen are combined to form helium-4, carbon-12, or oxygen-16, energy is also released. The former is exploited in an atomic bomb or a thermal nuclear reactor; the latter is used to produce hydrogen bombs and is currently being developed to create a fusion reactor. We will explore both systems in this volume.

Spurred on by Curie and Joliot's demonstration that radioactivity could be induced in previously stable nuclei by bombarding them with alpha particles Otto Hahn and Lise Meitner were joined in Berlin by Fritz Strassmann in 1934 to continue to

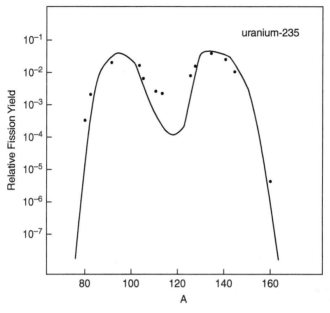

9. Fission fragment yields from uranium-235. The solid curve represents the fragment yields induced by slow, thermal neutrons. The points represent yields from fast neutrons (MeV) released in fission

investigate the phenomenon. Chadwick's 1932 discovery of the neutron led the Berlin group to investigate radioactivity induced by neutrons. Hahn and Strassmann found that uranium could be induced to undergo a new form of decay. Instead of simply spitting out small objects such as alpha or beta particles, the uranium split into two large fragments through the process of nuclear fission. Meitner and her nephew, Otto Frisch, quickly realised that fission was accompanied by a large release of energy. Enrico Fermi, who had recently left Europe for Columbia University, New York, then found that because the fission fragment isotopes have a smaller neutron excess than the heavier uranium (Figure 7), a number of neutrons were simultaneously emitted. The fission fragments were not of equal mass but fell in two groups centred around atomic mass numbers 95 and 140 (Figure 9).

10. Enrico Fermi, builder of the first atomic pile (nuclear reactor). Fermi was awarded the Nobel Prize in 1938 for his demonstrations of the existence of new radioactive elements produced by neutron irradiation and for his related discovery of nuclear reactions brought about by slow neutrons while working at the University of Rome. He was born in Italy in 1901 and died in the USA in 1954

In 1933, the Hungarian scientist Leo Szilard had speculated about the possibility of nuclear chain reactions and now Fermi was quick to see the possibility of realizing this with neutron-induced fission of uranium. He continued his studies at Columbia University with Szilard and developed the idea of a primitive self sustaining reactor. However, his early studies convinced him that such a device had to be much larger than his research facilities would allow.

The exigencies of the Second World War were about to change the whole pace of development.

Chapter 2

A new technology is developed

The first technology to be developed from the discoveries of nuclear physics was driven by the exigencies of the Second World War. For most non-scientists, including politicians, the laboratory experiments of Fermi, Curie and Joliot, and Chadwick appeared remote from the pressing issues of the approaching war. Amongst the scientific community, however, there was a growing speculation about the possibility of the explosive release of nuclear energy. Fermi's Columbia experiments had confirmed the concept of a chain reaction. Fears grew that Germany might seek to develop an atomic bomb.

Albert Einstein was one of very many European physicists who had fled the Continent ahead of the growing threat of National Socialism and the persecution of the Jews. Until this time, frontiers of scientific development had been concentrated in Europe and the Jewish community had played a leading role in cultural and scientific endeavours, especially in Germany. Of all the scientists in the world, Einstein had by far the highest popular profile. The European scientists brought with them to the USA the latest ideas about the new science of nuclear physics.

As concerns grew that Germany might develop a nuclear weapons programme, it was agreed that Einstein, although not directly involved in the evolving nuclear science, was the person most likely to be listened to by the politicians. In August 1939, a letter to President Franklin D. Roosevelt was drafted by Leo Szilard and signed by Einstein. The letter stressed the dangers that Germany might be able to develop a weapon of enormous destructive capacity. However, the letter was not delivered until October. In September, Hitler invaded Poland and the Second World War had begun. Despite the fact that the USA was not immediately involved in the war, the US government created the Uranium Committee which awarded a research contract with funding of $6,000. However, such was the fear of foreign scientists doing secret research that the money was not released to the Fermi–Szilard collaboration until Einstein was persuaded to send a second letter to the president in the spring of 1940.

With the new funding, Fermi was able to build the first atomic pile to reach criticality. The primitive reactor was built on the squash courts of the University of Chicago and went critical in December 1942. The Chicago Pile 1 was an essential development in the understanding of the fission process and the technical difficulties that would have to be overcome if nuclear technology was to deliver its potential.

The concept behind the atomic pile (the term 'nuclear reactor' was not coined until 1952) was simple: if neutron bombardment of uranium could induce fission with the release of a number of neutrons, these secondary neutrons could be used to induce further fissioning and a chain reaction could be established (Figure 11).

The problems were that the total available supply of uranium to the USA in the 1940s was limited and even this was of doubtful purity. This was a problem because the neutrons had to hit another uranium nucleus before they escaped from the reactor core. Thus a critical mass of uranium was required. Second, the most efficient fission process was to strike uranium-235 with slow-moving neutrons with a kinetic energy of less than 1 eV (see Figure 12). However, the secondary neutrons were emitted with much higher

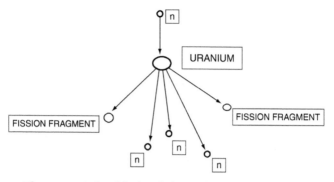

11. The neutron-induced fission chain reaction. A neutron strikes a uranium nucleus causing fission into two fragments with the release of several neutrons that can induce further fission

energies of a few MeV, and these were ineffective at causing fission. A material that moderates the velocity of the neutrons had to be found.

The induced fission yields are greater for uranium-235 at all neutron energies but at low energies the difference becomes dramatic, with the yield from uranium-238 becoming vanishingly small while it increases sharply for uranium-235 as the neutron energies are reduced.

When two particles collide, they transfer energy and momentum between themselves. In the case of billiard balls, the cue ball strikes a stationary target ball which takes energy and momentum from it and slows it down. If the target is much lighter than the projectile, the projectile sweeps it aside with little loss of energy and momentum. If the target is much heavier than the projectile, the projectile simply bounces off the target with little loss of energy. The maximum transfer of energy occurs when the target and the projectile have the same mass.

In trying to slow down the neutrons, we need to pass them through a moderator containing scattering centres of a similar

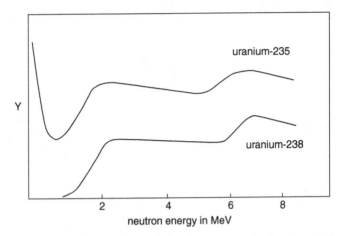

12. The probability of fission induced by neutron bombardment of uranium isotopes as a function of neutron energy

mass. The obvious candidate is hydrogen, in which the single proton of the nucleus is the particle closest in mass to the neutron. At first glance, it would appear that water, with its low cost and high hydrogen content, would be the ideal moderator. There is a problem, however. Slow neutrons can combine with protons to form an isotope of hydrogen, deuterium. This removes neutrons from the chain reaction. To overcome this, the uranium fuel has to be enriched by increasing the proportion of uranium-235; this is expensive and technically difficult. An alternative is to use heavy water, that is, water in which the hydrogen is replaced by deuterium. It is not quite as effective as a moderator but it does not absorb neutrons. Heavy water is more expensive and its production more technically demanding than natural water. Finally, graphite (carbon) has a mass of 12 and hence is less efficient requiring a larger reactor core, but it is inexpensive and easily available.

Another problem was that naturally occurring uranium is 99.3% uranium-238 and only 0.7% is in the fissile form uranium-235;

being chemically identical, no chemical process could separate them.

The first of the problems solved by the Fermi Pile was that of slowing down the secondary emitted neutrons from their initial MeV energies and thus increasing the fissionability. At the same time, the Pile was made large enough that there was a high probability that the neutrons would collide with a fissile nucleus before escaping from the Pile. Finally, a system was introduced to control the rate of nuclear reactions.

The initial sources of neutrons and the fuel for the Pile were pellets of natural uranium; these were separated by blocks of graphite which slowed down the neutrons as they passed through them. The assembly was roughly spherical and supported by a timber frame. The rate of the nuclear reaction was controlled by cadmium-coated rods (cadmium is a potent neutron absorber). Inserting the rods absorbed the neutrons and slowed down the fission rate. Removing the rods increased the number of neutrons until a self-sustaining nuclear chain reaction rate was achieved.

In Figure 13, a neutron strikes a uranium-235 nucleus and induces fission and the release of three neutrons. One of the neutrons escapes the core. Another is absorbed by a uranium-238 collision. The third collides with a uranium-235 nucleus, inducing fission and the release of two neutrons, both of which induce further fissioning, after being slowed by the graphite, in uranium-235 nuclei and the release of further neutrons. FFs are fission fragments.

Initially, there was little interest in the scientists' concerns in the USA. However, in the UK, two expatriate physicists, Otto Frisch and Rudolf Peierls, carried out a feasibility study of the possibility of fast fission of uranium-235 in 1940. This included an estimate of the amount of uranium required to create a critical mass and hence to make a bomb. The fast neutrons that were emitted during the fission process were most likely to strike the more abundant

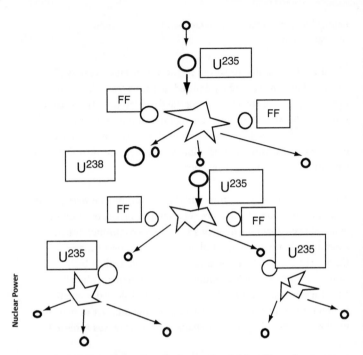

13. A schematic of a nuclear chain reaction initiated by a slow neutron collision with an isotope of uranium-235

uranium-238 isotopes. This produced little fission; however, it did result in the transmutation of uranium-238 into the first man-made element, plutonium-239, through the reaction

n + uranium-238 -> uranium-239 $-\beta^-->$ neptunium-239 $-\beta^-->$ plutonium-239

Plutonium-239 has a half-life of 24,000 years and does not exist in nature but lives more than long enough for any practical utilization. Plutonium had recently been identified by a team led by Glenn Seaborg at the University of California. By July 1941, the UK programme had shown that plutonium-239 was a much more potent fissile material than uranium-235.

Winston Churchill was so impressed by the Frisch–Peierls report that in September 1941 he authorized a programme to develop an atomic bomb. By December 1941, the USA had been drawn into the war, and in 1942 the Allies' nuclear research was coordinated with the title of the 'Manhattan Project', under the directorship of the American physicist Robert Oppenheimer. For security reasons, all work was now transferred to the USA. Also, it was quickly evident that under the impoverishment of war the UK could not match the resources available in the USA.

The work was divided over several sites in North America; Oak Ridge, Tennessee, was chosen as the facility to develop techniques for uranium enrichment (increasing the relative abundance of uranium-235), the Hanford site in Washington State became the centre for plutonium production, a number of Canadian sites produced heavy water, and the weapon design construction and testing were located at Los Alamos, New Mexico.

Since no chemical process could distinguish between uranium-235 and uranium-238, it required a physical process that depended on the mass difference, only 1.3%, to enrich uranium. At Oak Ridge, a giant gaseous diffusion facility was developed. Gaseous uranium hexafluoride was forced through a semi permeable membrane. The lighter isotopes passed through faster and at each pass through the membrane the uranium hexafluoride became more and more enriched. The technology is very energy consuming and Oak Ridge was chosen because the Tennessee Valley Authority had been established by President Roosevelt as a make-work programme during the Great Depression and this had created a gigantic hydroelectricity capacity. At its peak, Oak Ridge consumed more electricity than New York and Washington DC combined. Almost one-third of all enriched uranium is still produced by this now obsolete technology. The bulk of enriched uranium today is produced in high-speed centrifuges which require much less energy. The centrifuge facilities are composed of a large number of rotating cylinders. The gaseous uranium hexafluoride in the cylinders is subject to large centrifugal forces that throw the heavier isotopes to the outside leaving enriched uranium to be collected at the centres.

At Oak Ridge, Fermi was able to build a much larger version of his graphite pile and demonstrated that it was capable of producing plutonium. Following this demonstration, the plutonium production programme was transferred to Hanford. At Hanford, a giant version of the Fermi Pile was constructed specifically to produce plutonium. The reactor was fed enriched uranium from Oak Ridge and during its operation the fast neutrons transmuted some of the uranium-238 into plutonium-239. When the partially spent nuclear fuel was removed from the pile, the plutonium could be chemically separated from the uranium. Throughout the war years, the Manhattan Project followed the parallel approaches to producing weapon-grade material, uranium enriched to >98% uranium-235 and pure plutonium.

While the graphite-moderated Fermi Pile had demonstrated the possibility of a sustained nuclear chain reaction, graphite (carbon) was not the most efficient neutron moderator. Water, with its high hydrogen content, was a better moderator but this had the disadvantage that the hydrogen could capture the neutrons to form the heavy isotope deuterium. This had the effect of reducing the neutron flux necessary to maintain the chain reaction. There were two solutions to this problem. The loss of neutron flux could be compensated for by further enrichment of the uranium, or the water could be replaced with heavy water, i.e. water in which the normal hydrogen is replaced by deuterium. Given the limited capacity to enrich the uranium at the time, interest shifted to the use of heavy water. It was suspected that this was the approach being taken in Germany, and indeed this was the case. A contributory factor to the German failure to produce a nuclear weapon was their reliance on the extremely scarce resource of heavy water aggravated by the combined action of the UK air force and Norwegian resistance which destroyed a major heavy water production plant in occupied Norway. The French had removed the store of heavy water before Norway was occupied, and when the Germans sought to take the remaining heavy water to Germany after the destruction of the plant, the Norwegian resistance sank the ship carrying it. With hindsight, there was never any real chance that Germany could have produced an atomic bomb before the end of the war. Its technical options were limited, and it never had the resources – financial, manpower, material, and industrial capacity – to devote to a nuclear weapons programme on the scale of the Manhattan Project.

Natural hydrogen contains deuterium at the level of 1 part in 3,200. The difference in mass between the two isotopes means that at a given temperature molecules containing the isotopes are travelling at different velocities. In addition, chemical reactions involving molecules containing the different isotopes proceed at different rates. Canada took responsibility for producing heavy water as its contribution to the Manhattan Project and established a plant at Trail, British Columbia, in 1943, yielding six tonnes of 99% pure heavy water per year.

The Canadians used a two-step process: first, hydrogen sulphide gas and demineralized and deaerated water were circulated between a high-temperature (130°C) tower and a cold-water (30°C) tower. The gas and the water exchange hydrogen isotopes at a differential rate and a cascade process yields water with 15–20% heavy water content. The slightly enriched water is then fed into a cold distillation vessel consisting of a chamber in which the pressure is reduced to the vapour pressure of the water. This encourages evaporation, with the lighter molecules being given off preferentially. Repetition of the process can result in 99% pure heavy water.

In order to sustain a nuclear chain reaction, it is essential to have a critical mass of fissile material. This mass depends upon the fissile fuel being used and the topology of the structure containing it. For pure uranium-235, the critical mass of a sphere is 52kg with a radius 17cm. For pure plutonium, the corresponding figures are 10kg and 9.9cm. The chain reaction is maintained by the neutrons and many of these leave the surface without contributing to the reaction chain. Surrounding the fissile material with a blanket of neutron reflecting material, such as beryllium metal, will keep the neutrons in play and reduce the critical mass. Partially enriched uranium will have an increased critical mass and natural uranium (0.7% uranium-235) will not go critical at any mass without a moderator to increase the number of slow neutrons which are the dominant fission triggers. The critical mass can also be decreased by compressing the fissile material. An unmoderated supercritical mass of fissile material will build its spontaneous chain reaction into an explosion. In building the early atomic piles with relatively low enrichments of uranium and a relatively poor moderator (graphite), the structures at Hanford were enormous. This is not a great problem for a static ground level structure, but clearly it would be impossible to build a bomb of this size, nor, as was later to be the case, build a nuclear marine engine to power submarines and ice-breakers.

The Manhattan Project reached its final goal by developing two techniques for overcoming the size problem. First, using uranium from Oak Ridge enriched to greater than 99% uranium-235, a supercritical mass was created from two subcritical masses. One piece was shaped into a cylinder forming a 'target'. The other was shaped into a hollow cylinder 'bullet' into which the target would fit neatly to produce the supercritical mass. The bullet and the target were kept apart until a trigger of conventional explosive fired the bullet at the target and resulted in an atomic (*sic* nuclear) bomb. This weapon known as 'Little Boy' (Figure 14) destroyed the Japanese city of Hiroshima. The second system was used with nearly pure plutonium-239 from the Hanford facility. In this case, a sphere of plutonium encased a neutron source. The whole was surrounded by a blanket of conventional explosive which, when ignited, compressed the plutonium core with a spherical

28

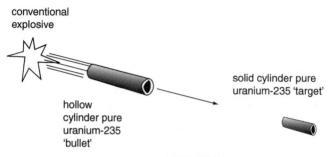

conventional
explosive

solid cylinder pure
uranium-235 'target'

hollow
cylinder pure
uranium-235
'bullet'

LITTLE BOY

14. A schematic of the operation of the Little Boy bomb

shockwave (Figure 15). Fat Man as this device was codenamed destroyed the city of Nagasaki and effectively brought the Second World War to an end.

With the war at an end, thoughts turned to the peaceful development of nuclear power. Unfortunately, the Cold War saw a continuation of the nuclear weapons race, with the USA and the USSR as the principal protagonists.

The end of the Second World War saw a schism develop between the Manhattan Project scientists. Many had joined the war effort specifically out of fear of Hitler's regime. With victory in Europe achieved, they were only too happy to return to their laboratories across the USA and Canada or to return home to the UK and Continental Europe. Many took up the cause of nuclear disarmament. Some, particularly those from Eastern Europe, saw in Stalin's USSR a threat just as great as that posed by Hitler. Prominent amongst these scientists was Edward Teller, who strongly advocated pressing on to develop a superbomb based not on the uranium/plutonium fission chain reactions but on the fusion of isotopes of hydrogen to form helium-4.

The principle of the hydrogen bomb was similar in some ways to the Nagasaki bomb. The plutonium core was replaced by a mixture

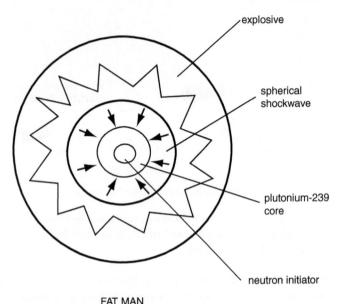

FAT MAN

15. A schematic of the operation of the Fat Man bomb. A conventional high explosive generates a spherical shockwave that compresses a core of pure plutonium-239 around a central neutron source

of the hydrogen isotopes deuterium and tritium seeded with lithium. To trigger the fusion chain reaction, the core had to be heated to temperatures and pressures greater than those at the centre of the Sun. Conventional explosives could not achieve this, but a plutonium bomb could.

Without the technical advances made during the Manhattan Project, in four short years, it is unlikely that a civil nuclear plant would have been constructed in the 20th century. Nor would the prospects of a fusion reactor, now predicted for the middle of the 21st century, appear likely before the 22nd.

Chapter 3
Thermal nuclear reactors

With the ending of the war, attention turned in many areas to the civil exploitation of technical advances made at such great speed by the exigencies of war; among these was the determination to turn the technology that had ended the war into a source of civil power.

It is now more than 50 years since operations of the first civil nuclear reactor began. In the intervening years, several hundred reactors have been operating, in total amounting to nearly 50 million hours of experience. This cumulative experience has led to significant advances in reactor design. Different reactor types are defined by their choice of fuel, moderator, control rods, and coolant systems. The major advances leading to greater efficiency, increased economy, and improved safety are referred to as 'generations'.

In 1954, the USSR connected an experimental reactor at Obninsk to the grid. The UK opened the first civil nuclear power station at Calder Hall in the north-west of England in August 1956. Calder Hall consisted of four Magnox reactors each generating 50MW of electricity. The Magnox reactors were based on the wartime graphite moderators and took their names from the magnesium non-oxidizing cladding of the fuel rods. As with all prototype devices, there was uncertainty about the length of time that the

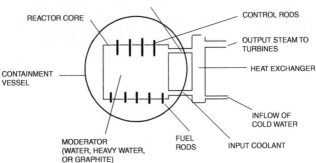

16. A graphic of the principal components of a thermal fission nuclear reactor. The control rods in this figure have been shown to be lowered into the core and the fuel rods inserted from below. Under gravity, the control rods would fall into the core and the fuel rods would drop out. This is a schematic of a 'passive' safety system

reactor could continue to operate, with 40 years an announced target. In fact, the reactor closed in 2003 having been operational for 47 years. A second plant at Chapelcross, in south-west Scotland, of similar design was connected to the grid in 1959.

While these reactors, like most of the first-generation reactors, were described to the public in terms of their electricity-generating capacity it is now known that they had the dual purpose to make electricity for public consumption and plutonium for the Cold War stockpiles of nuclear weapons. Many of the features of the design were incorporated to meet the need for plutonium production. These impacted on the electricity-generating cost and efficiency. The most important of these was the use of unenriched uranium due to the lack of large-scale enrichment plants in the UK, and the high uranium-238 content was helpful in the plutonium production but made the electricity generation less efficient. The government justified the increased cost by offsetting it by a notional value for the plutonium. However, no transfer of funding between the

Department of Defence and the electricity generators ever took place. In 1995, the UK government announced that all production of plutonium for weapons would cease. Today, these stockpiles of plutonium are a rich potential source of fuel for modern reactors.

The heat from the core of the Magnox reactors was transferred by the flow of high-pressure carbon dioxide gas. The first reactors had steel containment vessels while later ones used concrete. The hot gas then produced steam to drive the turbines that produced the electricity. The containment vessels are designed to ensure that whatever happens in the reactor core there is no significant release of radioactivity beyond the site boundary.

A number of problems were soon revealed. The high-pressure, high-temperature carbon dioxide was corrosive to mild steel. This led to the early reactors being run at a lower than optimal temperature further reducing their efficiency for electricity generation. Later plants used stainless steel and concrete. The need to refuel the reactor twice a year while the reactor was on load required sophisticated fuel-handling equipment. The Magnox alloy limited the maximum temperature and hence the thermal efficiency of the plant. It also reacted with water meaning that long-term storage of spent fuel under water was not possible. The need to reprocess fuel soon after its removal from the core greatly increased the hazards associated with the fission products resulting in the need for expensive remote handling equipment.

In all the UK built 11 Magnox reactors between 1956 and 1971. All bar one are now in decommissioning and the most recent, at Wylfa in Wales, ceased operation in 2010. At its peak, the UK's Magnox fleet generated approximately 4GW of power for the national grid. No two of these reactors were identical, hence each had to undergo separate licensing procedures. There were no economies of scale and thus costs were raised.

In 1961, a 161MW Magnox reactor was exported to Tokai Mura, Japan, and in 1963 a 153MW reactor to Latina in Italy. In 1986, when the design of the Magnox reactors was declassified, North Korea constructed a 5MW experimental plant at Yongbyon to produce plutonium for its nuclear weapons programme. Work on two further plants ceased in 1994 under the US–North Korean Agreed Framework.

In parallel with the UK, France built 9 graphite-moderated, carbon-dioxide-cooled nuclear plants between 1956 and 1972, and exported one to Vandellos, Spain. The major difference was that the Magnox fuel casings were replaced by a magnesium-zirconium alloy. The French reactors went under the title UNGG (Uranium Naturel Graphite Gaz). Prior to the development of its uranium enrichment facilities, Russia built 12 graphite-moderated reactors and 'exported' 4 to Ukraine. The technology behind these 'Generation I' civil power stations is now considered obsolete.

Like the UK, Canada had no access to uranium-enrichment facilities but did have experience of heavy water production. Canada has the second largest resource of proven uranium reserves in the world and has been involved in the development of nuclear technology since 1942, when a joint British–Canadian laboratory was established in Montreal. This soon moved to Chalk River, and in 1945 the Canadian Zero Energy Experimental Pile became the first source of a self sustaining nuclear reaction outside the USA. After the war, Canada established Atomic Energy of Canada Ltd to exploit its heavy water expertise to produce commercial reactors using unenriched uranium with a heavy water moderator and coolant.

Developed during the 1950s, the programme created the family of CANDU (Canada, Deuterium, Uranium) reactors that began operations in the 1960s. In total, Canada built 24 CANDU reactors (3 are currently being refurbished and 5 have been decommissioned). It has also exported 4 to South Korea, 2 to China, 2 to India, 1 to Argentina, and 1 to Pakistan. India has also built 13 reactors that are derivatives of the CANDU design and has a further 3 under construction. Canada ceased to cooperate with India in the development of nuclear technology in 1974, after India announced its independent nuclear weapons programmes. Relationships were resumed in 2009. Currently the most powerful of the CANDU installations is at Bruce River and this generates 6GW of electrical power; this is one of the largest nuclear power stations in the world.

The CANDU reactor is contained in a reinforced concrete building and the core of the reactor (calandria) is filled with heavy water under high pressure. Passing through the calandria there are a number of fuel rod assemblies and control rods. The fuel rod assemblies consist of Zircaloy tubes containing fuel pellets. These can be removed from one side of the calandria simultaneously with the insertion of a replacement assembly from the other side without taking the reactor off-line. The control rods are similar to those in the Magnox reactors. A secondary control system allows the injection of a neutron-absorbing, high-pressure gadolinium nitrate solution into the moderator. The heavy water also serves as the coolant transferring heat to light water which turns into steam to drive the turbines.

An advantage of the CANDU system is that it can deal with a range of fuels. Initially designed for natural uranium, it can also operate with enriched uranium and thorium-233 (of great interest to India, which has large reserves of thorium). A disadvantage is that the heavy water moderator under neutron bombardment produces small quantities of the radioactive isotope tritium (hydrogen-3). This can produce very heavy water which has to be kept from escaping into general water supplies. Tritium is also a primary ingredient in the production of hydrogen bombs.

Meanwhile, the USA had access to specialized plutonium production facilities at Hanford supplemented by supplies from the UK. The USA also had extensive uranium enrichment facilities and could afford to look at the possibility of using natural water as a moderator and coolant. The fact that the water absorbed neutrons could be compensated for by increasing the fraction of fissile uranium-235.

In 1954, a military nuclear plant at Fort Belvoir, Virginia, became the first US reactor to supply electricity to a commercial grid. In the same year, the first nuclear powered submarine, the *USS Nautilus*, was launched.

The USA concentrated its efforts on the development of PWRs (Pressurized Water Reactors). The pressurization of fluids raises their boiling point; in the case of PWRs to well above the boiling point of water at atmospheric pressure. These reactors consisted of a pressure vessel containing a core of fuel rods, control rods, and high-pressure natural water as a moderator. The moderator also acts as the coolant but becomes radioactive during operation and is circulated through a heat exchanger linked to an external, non-radioactive supply of water that turns to steam and drives the turbines.

A variation on the PWR is the BWR (Boiling Water Reactor). This is similar to the PWR but is operated at a lower pressure, allowing the coolant to turn to steam to drive the turbines directly. The BWRs are inherently more thermally efficient and more compact than the PWRs.

In the 1960s, Russia developed its own version of the PWR known as VVER (Voda-Voda Energetichesky Reactor, water-moderated, water-cooled energetic reactor). Compact versions of the VVER are also used to power Russia's considerable nuclear fleet.

Generically, the PWRs, BWRs, and VVERs are known as LWRs (Light Water Reactors). LWRs dominate the world's nuclear power programme, with the USA operating 69 PWRs and 35 BWRs; Japan operates 63 LWRs, the bulk of which are BWRs; and France has 59 PWRs. Between them, these three countries generate 56% of the world's nuclear power. In terms of fractions of total electricity generation, France leads, with 80% of its electricity coming from nuclear plant.

In the UK, experience with the Magnox reactors gave rise to the development of AGRs (Advanced Gas-Cooled Reactors). The AGR concept was similar to that for the Magnox plants, but the design allowed much higher gas temperatures and hence greater thermal efficiency. The higher temperatures required the Magnox fuel cladding to be replaced with stainless steel. This in turn produced greater neutron absorption in the cladding, leading to the need to use enriched uranium. In total, seven nuclear plants were constructed and are currently operating in the UK. Each plant consists of two AGRs.

The PWRs, BWRs, VVERs, CANDUs, and AGRs are referred to as Generation II nuclear reactors.

Modifications of the Generation II reactors have led to the design of an advanced BWR by the General Electric Company. This is a BWR with an improved circulation system, with the pumps inside the pressure vessel. The absence of external pumping and pipe-work reduces the capital cost, increases safety and thermal efficiency, and lowers operating costs. Four ABWRs are operating in Japan. Mitsubishi have increased the efficiency of the PWR by incorporating neutron reflectors. The APR design is currently awaiting certification. It is likely the ABWR and APWR Generation III reactors will be superseded by more recent designs.

In the 1990s, a series of advanced versions of the Generation II and III reactors began to receive certification. These included the

ACR (Advanced CANDU Reactor), the EPR (European Pressurized Reactor), and Westinghouse AP1000 and APR1400 reactors (all developments of the PWR) and ESBWR (a development of the BWR). Orders for the Generation III+ reactors are now being placed and some construction has begun, with the first plants to become operational in 2012.

The ACR uses slightly enriched uranium and a light water coolant, allowing the core to be halved in size for the same power output. AECL hope to have the first ACR on line by 2016. The ESBWR is General Electric's development of the APWR. The modified design allows for natural circulation by reducing the length of the core. The lack of pumps reduces the capital cost, increases the reliability and reduces the running costs of the plant. It would appear that two of the Generation III+ reactors, the EPR (marketed in Europe as the European Pressurized Reactor, and globally as the Evolutionary Power Reactor) and AP1000, are set to dominate the world market for the next 20 years.

The EPR is produced by the world's largest reactor manufacturer Areva (a French–German consortium). The EPR is a development of the PWR designed to use highly enriched uranium oxide fuel or a mixture of uranium and plutonium oxide, known as MOX. Compared with the earlier PWRs, the EPR is much more compact and hence involves much lower construction costs. The EPR has a number of enhanced safety features, such as four physically separate emergency systems, each capable of performing 100% of the safety function on its own. Two of the systems are resistant to direct aeroplane crashes. There is a leak-tight containment vessel capable of withstanding high temperatures and pressures, and containing radioactive material in the case of severe accidents leading to core meltdown. In the highly unlikely event of molten core material escaping from the reactor vessel, there is a passive collection and cooling system. The construction is particularly robust and designed to protect the installation from direct large civilian aeroplane crashes and severe earthquakes. The Union of Concerned Scientists has stated that the EPR is considerably safer than current reactor designs.

Construction began in 2005 of the first EPR at Olkiluoto, Finland. Originally planned to open in 2009, construction delays have put back the expected start-up until 2012; the same year that a second EPR plant at Flamanville, France, is scheduled to start operations. An additional French plant at Pleny is planned for construction starting in 2012 and completion by 2017. In 2007, China placed an order for two EPRs, to be built at Guangdong, with options for a further two.

In 2008, the French electricity utility EDF, a partner in Areva, bought British Energy, the operator of the UK's fleet of AGRs. It is likely that a number of EPR and AP1000 reactor stations will be built in the UK on the current sites of Magnox and AGR plants facing decommissioning. In November 2009, the UK government signalled its wish to have ten new nuclear reactors operating by 2022. It simultaneously announced a streamlining of the planning and licensing procedures to fast-track the process. Following a change of government in 2010, doubts have been raised about the timing of the start of construction of these plants, and it would appear that the target may have been reduced to four, the first two of which are to be built at Hinkley Point by 2020.

In 2009, India contracted to build two EPR plants at Maharashtra, and France agreed to supply four plants to Italy, the first to be operational by 2020. In the USA, there are a number of applications for combined construction and operation licences awaiting final approval of the EPR design by the NRC (Nuclear Regulatory Commission) that are due imminently (2009).

The AP1000 is produced by the Westinghouse Company which was previously owned by BNFL (British Nuclear Fuels) and was sold to the Japanese company Toshiba in 2006. The AP1000 uses passive safety systems, that is, relying on gravity and natural circulation rather than pumps. As a result, like the EPR, the 'plumbing' of the AP1000 is greatly simplified compared with earlier PWR designs. This results in a halving of the amount of pumps, pipe-work, and buildings. The modular structure of the design is intended to reduce construction time to three years. These features combine to lead to a considerable capital cost saving. A major advance is that the Generation III+ reactors produce only about 10% of waste compared with earlier versions of LWRs. The subject of nuclear waste will be considered in Chapter 4.

At the end of 2005, the US NRC approved the design of the AP1000, opening the way for construction applications from US utilities. The first AP1000 units will be built in China. In February 2008, construction started on a four-unit plant at Zhejiang for completion by 2013–15. In July 2008, construction began on a two-unit plant at Shandong for completion in 2014–16. China has officially adopted the AP1000 design as a standard for future nuclear plants and has indicated a wish to see 100 nuclear plants under construction or in operation by 2020.

In the USA (2009), six applications for combined construction and operating licences have been filed, each for two reactors. The first two AP1000s in the USA will be built at the Vogtle Electric Generating Plant in Georgia. These will be the first new nuclear plants in the USA since the 1979 Three Mile Island incident (see Chapter 5).

The Generation I reactors were designed to produce plutonium which was extracted for military purposes. Even in the conventional thermal nuclear reactors we have been discussing, much of the heat released in the chain reaction comes from the fission of plutonium at later stages of the fuel cycle. This then holds out the possibility of building a nuclear reactor that actually breeds more fissile fuel than it consumes. The production of plutonium-239 comes from the fast neutron collisions with uranium-238 nuclei, and such reactors are designated FBRs (Fast-Breeder Reactors). The USA, UK, France, Russia, India, and Japan have built developmental FBRs. Since FBRs are generating fissile fuel during their operating cycle, it is desirable to co-locate fuel reprocessing facilities on site. The US Integral Fast-Breeder Reactor had an on-site fuel reprocessing unit that recycled all uranium and transuranic elements, including plutonium, via electroplating. This left just short half-life fission products as waste which could be separated to provide isotopes of medical or industrial use with only a tiny residue of waste sent to the waste repository.

FBRs use greatly enriched fuel; frequently a mixture of uranium dioxide and plutonium dioxide. The core may be surrounded by a blanket of uranium in which fast neutrons escaping the core can breed more plutonium, or the blanket may be composed of a neutron reflector. No moderator is required.

The cores are much more compact than in conventional nuclear reactors and hence the operating core temperatures can be higher. To carry away the heat sufficiently rapidly would require an enormous through-put of water in the core. The neutron absorption by water would require even greater fuel enrichment. To date, all FBRs have had liquid metal coolants. There have been some prototypes using mercury, lead, and a sodium potassium alloy as coolants, but all large-scale reactors have used liquid sodium. There are two basic coolant systems: in the 'loop' system, the primary coolant circulates through primary heat exchangers external to the reactor tank. The primary heat exchanger must be within the reactor's biological shield because of the presence of radioactive sodium. In the 'pool' system, the primary heat exchanger and circulators are immersed within the reactor tank.

The first US FBR went critical in 1951 at Idaho Falls. It led to the development of a second experimental reactor that operated between 1964 and 1994 which was designed to study on-site fuel recycling. The only commercial station built in the USA was at Lagoona Beach, Michigan, and went into operation in 1956. It was closed in 1966 when a fault led to a partial meltdown of the core.

In the UK, the Dounreay Fast Reactor went critical in 1959 and led to the development of the Prototype Fast Reactor which connected to the grid in the 1970s. Again, at Dounreay a central part of the programme was to investigate the on-site fabrication and recycling of fuel; the programme closed in 1994. France's first FBR was named Rapsodie and went critical in 1976 near Aix en Provence and was used to study the on-site reprocessing of nuclear waste. In 1984, the largest FBR to be built to date entered service. The Superphénix was closed in 1997 on economic and political grounds.

The USSR built a number of small experimental FBRs between 1955 and 1969. The first full-scale power reactor was built in Kazakhstan in 1973. Rated at 350MW, this reactor delivered 130MW to the local grid and was used as a desalination plant, delivering 80,000 tonnes per day of fresh water to the city of Aktau. At the time of the break-up of the USSR, plans were advanced for 800MW and 1600MW fast-breeder reactors. The 800MW reactor is still in operation and the 1600MW reactor is scheduled to come on line in 2016.

Japan built two fast-breeder demonstration reactors. Construction of the second reactor began in 1985 and it achieved criticality in 1994, but was closed in 1995 as the result of a fire. The Japanese government has vested Mitsubishi Heavy Industries with responsibility for the future development of FBR technology.

India has developed the technology to produce a plutonium-rich uranium-plutonium mixed carbide fuel for its FBR programme. India's first fast-breeder test reactor went critical in 1985. Work has begun on a 500MW prototype fast-breeder reactor near Chennai. India is seeking to exploit its rich reserves of thorium to extract nuclear fuels and has a strong development programme based upon the use of thorium-232 as a fuel in a thermal breeder reactor and has announced its intention to build four more 500MW FBRs.

The uranium cycle is based on the slow neutron fission of uranium-235 and the fast neutron transmutation of uranium-238 into plutonium-239 which, like uranium-235, fissions under neutron collisions. The thorium cycle is based on the plentiful supply of thorium-232, which under fast neutron bombardment is transmuted into uranium-233, which is again fissile. Uranium-233 has a half-life of 160,000 years and does not occur in nature. However, it exists long enough in a reactor core to be a fissile fuel.

At first, it was believed that the future of nuclear power rested on the programme to develop the FBR. However, the low cost of enriched uranium and plutonium from decommissioned weapons makes the FBR technology uncompetitive compared to the new Generation III+ reactors. Several countries (the USA, UK, France, and Russia) have cancelled or scaled down their efforts to develop FBRs.

While globally the development of FBRs has stalled, in part due to the low-cost supply of plutonium from decommissioned weapons, the technical advances made in dealing with the problems they posed has helped with the development of Generation IV reactors. These comprise a series of designs currently being researched. An international forum has been established to monitor the development of these reactors which are being designed to improve safety, proliferation resistance, waste management, natural source utilization, and to reduce the capital and running costs of nuclear plant. The Generation IV reactors are principally of two types: there are VHTRs (Very High Temperature Reactors) and FRs (Fast Reactors, not designed to breed fissile fuels).

The VHTRs are graphite-moderated and uranium fuelled but designed to operate at much higher temperatures than Generation II and III reactors. With outlet temperatures of 1000°C, they would have a higher thermal efficiency, at around 50% compared with earlier designs at 33%. The first VHTRs are expected to go on line in the early 2020s.

A by-product of VHTR operations is the use of waste heat to generate hydrogen by means of the sulphur–iodine cycle. This is a series of chemical reactions in which iodine and sulphur dioxide act as catalysts with water and heat to produce hydrogen and oxygen.

A novel version of the VHTR is the 'pebble bed' design. This has fuel elements composed of pyrolitic graphite impregnated with micro-particles of fissile fuel and encased in a ceramic coat which provides structural integrity and contains the fission products. Thus each fuel element, about the size of a tennis ball, carries its own moderator. The fuel elements are called 'pebbles'. To create a critical mass, 360,000 pebbles are amassed to form the core. The coolant is an inert gas like helium or a semi-inert gas like nitrogen or carbon dioxide. The design of Pebble Bed Reactors originated in Germany and for several years was actively pursued in South Africa. A prototype Pebble Bed Reactor, the HTR-10, is currently operating in China.

The Pebble Bed Reactor design is extremely flexible. Increasing power capacity can be built up incrementally by adding modules. Reactors with as little as 1.5MW capacities have been proposed to provide marine engine power.

As we discussed with FBRs, large amounts of water in the core inhibit the nuclear reaction by absorbing neutrons. However, supercritical water, that is, water under pressure heated to temperatures far in excess of the boiling point of water at atmospheric pressure, has a sufficiently high heat capacity that it can be used in the reactor core in much smaller quantities. This has led to the concept of the SCWR, the Supercritical-Water-Cooled Reactor. Basically, this is a PWR operating at much higher pressures. The working fluid can be generated in conventional supercritical fossil-fuelled boilers and it is hoped that this will reduce the cost of electricity generation.

Other VHTR designs envisage the use of molten salt or sodium as coolants, and many are interested in using the thorium–uranium fuel cycle instead of the current uranium–plutonium system.

There are a range of FR designs under investigation. All have closed fuel cycles and fast neutrons. This allows uranium, together with plutonium and other transuranics that are produced, to be consumed in the core, resulting in their being no need for long-lived radioactive products to leave the site. The SFR (Sodium Fast Reactor) draws on experience of metal-cooled FBRs combined with closed fuel-cycle technology. The GFR (Gas-Cooled Fast Reactor) is helium-cooled and is being used to investigate the use of ceramic-clad fuel pellets to contain fission products. The LFR (Lead-Cooled Fast Reactor) has liquid lead or lead/bismuth as a coolant. Modular structures allow plants ranging from 50MW to 1,200MW ratings. The LFR is cooled by natural convection, with outlet coolant temperatures of 550°C; it is hoped to raise this to 800°C with the use of advanced materials. It is unlikely that fully commercial Fast Reactors will be available before the 2030s.

All the Generation IV reactors claim to be inherently safe and to produce much less waste than current designs. The retention of fissile fuel within the core reduces the opportunities for removal of weapon-grade material and hence the risk of weapons proliferation. The higher operating temperatures mean greater thermal efficiency. The use of the thorium–uranium cycle greatly increases global reserves of nuclear fuel. There is approximately four times more thorium on Earth as there is uranium. 100% of all naturally occurring thorium is the isotope thorium-232, the seed corn for the thorium-232 to uranium-233 fast neutron-induced transmutation; whereas for uranium-fuelled reactors, only 0.7% of the naturally occurring uranium is the fissile uranium-235.

All thermal electricity-generating systems are examples of heat engines. A heat engine takes energy from a high-temperature environment to a low-temperature environment and in the process converts some of the energy into mechanical work. In the case of the PWRs, for example, the temperature across the core is 275–315°C. The superheated steam that drives the turbines is about 275°C. To produce this supercriticality, the pressure is maintained at around 60 times atmospheric pressure. In general, the efficiency of the thermal cycle increases as the temperature difference between the low-temperature environment and the high-temperature environment increases. In PWRs, and nearly all thermal electricity-generating plants, the efficiency of the thermal cycle is 30–35%. At the much higher operating temperatures of Generation IV reactors, typically 850–1000°C, it is hoped to increase this to 45–50%.

During the operation of a thermal nuclear reactor, there can be a build-up of fission products known as reactor poisons. These are materials with a large capacity to absorb neutrons and this can slow down the chain reaction; in extremes, it can lead to a complete close-down. Two important poisons are xenon-135 and samarium-149. Most of the xenon arises from the decay of the fission product iodine-135 (half-life 6.6 hours). During steady state operation, the xenon builds up to an equilibrium level in 40–50 hours when a balance is reached between the rate of iodine production, and hence of xenon, and the burn-up of xenon by neutron capture. If the power of the reactor is increased, the amount of xenon increases to a higher equilibrium and the process is reversed if the power is reduced. If the reactor is shut down the burn-up of xenon ceases, but the build-up of xenon continues from the decay of iodine. Restarting the reactor is impeded by the higher level of xenon poisoning. Hence it is desirable to keep reactors running at full capacity as long as possible and to have the capacity to reload fuel while the reactor is on line. The samarium is not radioactive and the concentration builds to an equilibrium value in around 500 hours.

Electricity demand varies with the seasons and with the time of day. An efficient generating mix is required that can respond to these timescales for change. The minimum requirement in the temperate zone occurs in the early hours of the morning in the summertime. This minimum level of electricity generation is known as the 'base load'. Nuclear plants operate at highest efficiency when operated continually close to maximum generating capacity. They are thus ideal for provision of base load. If their output is significantly reduced, then the build-up of reactor poisons can impact on their efficiency.

There are various other fission product poisons which, although individually insignificant, collectively are important. Frequently, the reactor needs a clean out of poisons when only 3–5% of the fissile fuel has been used. This makes fuel reprocessing an important part of the nuclear cycle; a subject we will deal with in the next chapter.

Reactor poisons can be deliberately introduced into the reactor as part of the control and safety system. The primary control system usually consists of rods of boron steel that can be moved in and out of the core to control the rate of the chain reaction. Additional control can be exercised in water-cooled reactors by adding soluble poisons like boric acid, sodium polyborate, or gadolinium nitrate to the coolant in the case of an emergency shut-down.

The operational lifetime of reactors is uncertain. Generation II and III reactors were designed to operate for about 40 years, and many have done so and are currently being given extensions to their licence on an *ad hoc* basis following safety checks. An example of the sort of restriction to a reactor's life is provided by the graphite-moderated reactors. For safe and efficient operation, the fuel rods and control rods have to move freely through channels in the graphite core. At the high operating temperatures and radiation levels, the graphite degenerates and channels can become blocked and it becomes unsafe to continue to employ it. As channels close, the efficiency of the reactor is reduced finally to the point where decommissioning becomes necessary. Currently the UK's AGR fleet are receiving limited extensions of 5 to 10 years, but many may not remain operational for the full period. PWRs face different problems and are currently receiving extensions of licence of 10 to 20 years. Many of the currently operating reactors were built in the late 1960s and 1970s. With a global hiatus on nuclear reactor construction following the Three Mile Island incident and the Chernobyl disaster, there is a dearth of nuclear power replacement capacity as the present fleet faces decommissioning.

Nuclear power stations, like coal-, gas-, and oil-fired stations, produce heat to generate electricity and all require water for cooling. The US Geological Survey estimates that this use of water for cooling power stations accounts for over 3% of all water consumption. Most nuclear power plants are built close to the sea so that the ocean can be used as a heat dump. There are reports that this warming of the water has resulted in increased scallop harvests, and anglers in the Finger Lakes of upper New York State reported larger fish sizes following the construction of a nuclear power station. In Table 1, we present typical water consumption rates for various power-generation technologies. The need for such large quantities of water inhibits the use of nuclear power in arid regions of the world.

As a footnote to the discussion of Generation I–IV reactors, I should note the existence of Generation 0 reactors. Today, the low relative abundance of uranium-235 means that in the absence of a moderator there is no critical mass for natural

HOURLY COOLING WATER CONSUMPTION IN LITRES PER MW

NUCLEAR	3080
COAL	1760
GAS	660
HYDRO	6160
SOLAR THERMAL	4400
GEOTHERMAL	8800
BIOMASS	1760

Table 1. Water usage in various means of power production. The bulk of water consumption for hydroelectricity comes from reservoir evaporation and seepage into the water table. The higher the operating temperature, the greater the water usage. Since large coal, gas, and nuclear plants are usually rated at around a GW, they can consume millions of litres per hour

uranium. With natural water as a moderator it is necessary to enrich the uranium to a 3–5% abundance of uranium-235. It was not always so.

Uranium-238 has a half-life of 4.5 billion years, comparable to the age of the solar system. Uranium-235 has a half-life of some 700 million years and is hence decaying at a faster rate. If we go back in time 2 billion years, the relative abundance of uranium-235 was 3–4%, and with water and the right geological geometry it was possible for a natural thermal reactor core to be established. The first such Generation 0 reactor was discovered in the Oklo Valley, Gabon, West Africa, in 1972. Studies of the relic isotope abundances give much information about the operating characteristics of this natural reactor. The chain reaction was initiated when a uranium-rich mineral deposit was inundated with natural ground water. The heat generated by the nuclear reactions boiled off the water and the chain reaction ceased and fission product poisons decayed. When water came back, the cycle started again and continued for hundreds of thousands of years until the chain reaction could no longer be supported.

An interesting question arises as to why there was no earlier natural reactor given the even greater uranium enrichment. Uranium metal is not water soluble, whereas uranium oxide is. Solutions of uranium oxide can be transported by the flow of water. Deposits from these water flows can build up rich mineral deposits sufficient to achieve a critical mass. Uranium oxide only formed in significant quantities once the Earth's atmosphere became oxygen rich some 2 billion years ago.

Chapter 4
Nuclear fuel reprocessing and radioactive waste

A nuclear reactor is utilizing the strength of the force between nucleons while hydrocarbon burning is relying on the chemical bonding between molecules. Since the nuclear bonding is of the order of a million times stronger than the chemical bonding, the mass of hydrocarbon fuel necessary to produce a given amount of energy is about a million times greater than the equivalent mass of nuclear fuel. Thus, while a coal station might burn millions of tonnes of coal per year, a nuclear station with the same power output might consume a few tonnes.

The nuclear fuel is inserted into the reactor core in containment rods which themselves become radioactive under the intense radiation to which they are exposed. In addition, some reactor cycles expose coolants like water or inert gasses to radiation.

There are a number of reasons why one might wish to reprocess the spent nuclear fuel. These include: to produce plutonium either for nuclear weapons or, increasingly, as a fuel-component for fast reactors; the recycling of all actinides for fast-breeder reactors, closing the nuclear fuel cycle, greatly increasing the energy extracted from natural uranium; the recycling of plutonium in order to produce mixed oxide fuels for thermal reactors; recovering enriched uranium from spent fuel to be recycled through thermal reactors; to extract expensive isotopes which are of value to medicine, agriculture, and industry. An integral part of this process is the management of the radioactive waste.

Currently 40% of all nuclear fuel is obtained by reprocessing. The first nuclear reprocessing system was developed at Oak Ridge in 1943 as part of the Manhattan Project. This was restricted to proving the concept of producing weapon-grade material, pure plutonium, from spent natural uranium and fission products. The process successfully yielded a few grams of plutonium. As we have seen, towards the end of 1944 the plutonium extraction process moved to the Hanford site and was scaled up for large-scale production. The original Oak Ridge bismuth phosphate process could only extract plutonium. In 1949, Oak Ridge developed a solvent process that yielded both pure uranium and pure plutonium. A further separation plant was built at the Savannah River site.

Following India's demonstration that an independent nuclear weapons programme could be based on recycled waste from a commercial reactor, the USA banned all commercial reprocessing and plutonium extraction as a result of concerns about nuclear weapon proliferation. The ban was lifted in 1981, but it was not until 1999 that the US Department of Energy placed a contract with a consortium of US companies and Areva (the world's largest nuclear processor) to design and construct a facility for fabricating mixed oxide fuels; construction work began in 2005.

In 2006, the US Department of Energy proposed the formation of a Global Nuclear Partnership to promote the use of nuclear power and close the fuel cycle in order to reduce waste and the risk of weapon proliferation. Effectively the proposal would divide the world into nuclear power user nations and fuel provider nations. The provider nations would supply enriched uranium and take back spent fuel for reprocessing. As of 2008, 25 nations had joined as full members of the partnership and a further 17 had observer status. Notable amongst those who have not joined are India, Iran,

North Korea, and Pakistan, all of whom have demonstrated nuclear weapon-building capacity. Among others that are not fully signed-up members are Brazil, Israel, and South Africa, often rumoured to be close to an independent nuclear capacity. South Africa, one of the largest sources of uranium, has stated that 'to mine uranium for export only to buy it back in enriched form would not be in our national interest'. An essential part of the programme was the production of plutonium from spent fuel in a form that allows it to be used as a reactor fuel but rendered it unsuitable for weapons. In July 2009, the project was cancelled on the orders of President Obama.

In the late 1940s and early 1950s, the UK, USSR, and France were developing separate nuclear weapons capability alongside civil nuclear power programmes incurring the need for reprocessing facilities.

In 1947, the UK took the decision to develop a separate nuclear weapons programme in close collaboration with the USA. An atomic pile was built at Windscale, Cumbria (since renamed Sellafield), to produce plutonium for its weapons programme and for exchange with the USA. The pile went critical in 1950 and by 1952 plutonium, uranium, and fission products were being separately extracted. The uranium was recycled as fuel for the growing fleet of Magnox reactors and plutonium was being produced at a rate of 300 tonnes per year by 1961, some of which was used as fuel for the Dounreay FBR. Dounreay also had on site reprocessing capacity for its closed fuel cycle operations until it closed in 1994. In 1990, a second reprocessing plant known as THORP (Thermal Oxide Reprocessing Plant) was opened at Sellafield with a capacity to deal with 1,600 tonnes of spent LWR fuel per year.

Between 1945 and 1948, the USSR built a nuclear facility at Mayak. This consisted of five atomic piles and reprocessing facilities to extract plutonium for its weapons programme.

The Mayak facility has been plagued with incidents leading to workers and the local population receiving high levels of radiation. Today the plant specializes in the production of tritium and applicable radio-isotopes and the processing of plutonium from decommissioned weapons. Russia is constructing a reprocessing plant for spent fuel from VVERs at Zheleznogorsk.

In 1960, France became the fourth nation to explode an atomic bomb. French reprocessing capacity is concentrated at the La Hague site, on the Contentin Peninsula, owned and operated by Areva. The La Hague site is the largest reprocessing site in the world, with over half the global capacity at 1,700 tonnes of spent fuel per year. It is designed to handle waste from LWRs. To date over 10,000 tonnes of spent fuel have been reprocessed. The plant handles spent nuclear fuel from France, Japan, Germany, Belgium, Switzerland, Italy, and the Netherlands. The spent fuel arriving at the plant consists of 96% uranium, 1% plutonium, and 3% non-reusable final waste (raffinate). The uranium and plutonium are recycled and the non-renewable final waste is packaged and returned to the country of origin as required by international law. France also has a small plant at Marcoule for reprocessing waste and fabricating mixed oxide fuels for its experimental FBR programme and Generation III+ reactors.

India has two small military reprocessing facilities, at Kalpakkan and Trombay, and a civil facility at Tarapur for dealing with waste from its CANDU-type reactors. Japan has two reprocessing facilities, at Tokaimura and Rokkabasho, for handling LWR spent fuel.

The original Oak Ridge system for extracting plutonium, known as the Bismuth Phosphate Process, is now considered obsolete. The spent fuel came in the form of radium and reactor products clad in aluminium. First, the aluminium was stripped off by boiling it in caustic soda. The spent fuel was then dissolved in nitric acid. The addition of bismuth nitrate and phosphoric acid led to the co-precipitation of plutonium and bismuth phosphate as solids leaving the fluid containing the spent uranium and fission products. Potassium permanganate was then added to hold the plutonium as a highly ionized form of plutonium dioxide. The bismuth phosphate was then removed by re-precipitation, leaving the plutonium in a solution. Following further chemical treatment, the plutonium was then extracted in the form of plutonium nitrate.

While the pure plutonium was required for weapon material, this represented only 1% of the spent fuel. The uranium that comprised 96% was treated as waste along with the fission products and discarded. Civil spent fuel reprocessing requires the extraction of both the plutonium and uranium to close the fuel cycle.

The current standard method for extracting both uranium and plutonium separately is the PUREX (Plutonium and Uranium Recovery by Extraction) process. This is a liquid-liquid extraction process that separates chemical compounds based upon their relative solubilities in two immiscible liquids, i.e. two liquids that do not form a compound (oil and water). When used with long-cycle spent reactor fuel, the presence of too much plutonium-240 renders the extracted plutonium unsuitable as a weapon-grade material (only the plutonium-239 is fissile). With shorter fuel cycles, weapon-grade plutonium can be extracted. Because of this, PUREX chemicals are monitored either by national agencies or the IAEA (International Atomic Energy Agency).

A development of the PUREX process is the UREX process which only extracts the uranium in an effort to reduce the risk of nuclear weapon proliferation. This is achieved by the addition of acetohydroxamic acid at the extraction phase; this locks in the plutonium.

The PUREX process can be augmented by various systems to extract other radioisotopes from the residual waste fluid. The US

TRUEX process is designed to remove the transuranic metals in order to reduce the alpha radioactivity of the residual waste. The SANEX (Selective Actinide Extraction) is designed to extract lanthanides and actinides. The lanthanides are a serious reactor poison and must be removed if the actinides are to be recycled as fuel or used in other industrial processes. The Russian UNEX process aims to remove the most dangerous radioisotopes of strontium and caesium and other minor actinides. In addition, nuclear waste contains costly noble metals which can be extracted.

When it comes to the cost of reprocessing, the economics depends on the objectives. If we consider the simplest situation where the waste from a thermal reactor is reprocessed for a single recycling of plutonium and compare it with an open cycle and direct disposal of waste, then the recycling is, at present, an expensive option. However in such an open cycle we have a considerable quantity of highly radioactive material requiring long-term secure storage both for protection against the radiation hazard and its possible theft to fuel nuclear weapon proliferation. If the uranium is extracted and recycled, the amount of direct waste is reduced by 96%. Using MOX fuels uses plutonium in a thermal reactor and hence denies its availability for proliferation. This is particularly important given the large stockpiles of weapon-grade plutonium accumulated during the Cold War.

If the waste is simply left to radioactively decay, then after 40 years its radioactivity has dropped by 99.9%, but it is over a thousand years before it falls to the level of naturally occurring uranium. The most dangerous long-lived components of the waste are plutonium and the other transuranics, for which the activity remains high for more than 10,000 years unless reused as nuclear fuel.

Reprocessing the spent fuel and separating the various components reduces, greatly, the bulk of material requiring long-term secure disposal and allows specialized storage or deactivization techniques for the remainder.

So far we have discussed radioactive waste from fuel recycling. With roughly 3,000 tonnes of fuel recycled each year and 3% of this unusable waste we are faced with disposal of some 90 tons globally per year. The amount is actually much greater than this due to contamination of the solvents and other chemicals used in the reprocessing.

The list of all waste associated with the nuclear power industry includes: uranium mine tailings, uranium (both natural and reprocessed) and its decay products (both fission products and suburanium radioactive chain products), the transuranic elements (plutonium and minor actinides), together with a range of activation waste (materials that were not initially radioactive but have become so as a result of irradiation, e.g. fuel-rod cladding). The waste is classified as low-level waste (LLW), intermediate-level waste (ILW), and high-level waste (HLW), and its disposition depends on its level and composition.

Uranium tailings are very LLW but contain hazardous heavy metals such as lead and arsenic. To date, as in many other mining industries, no great efforts have been made to manage this waste.

Most LLW does not come from the nuclear power industry *per se*, but from hospitals and other industries. Radioactive isotopes are used as tracers in medicine, agriculture, and many industries. They are frequently used in smoke detectors. They are also used in radiotherapy to treat cancer. Often, the LLW designation is used as a precautionary label even when there is little hazard associated with it. It is frequently compacted or incinerated in order to reduce its bulk before it is disposed of in shallow land burial; it usually contains small traces of short-lived isotopes.

The ILW label is not recognized in the USA but is used in Europe and other parts of the world. It consists of materials such as activation waste, e.g. fuel-rod cladding and some contaminated decommissioning material. The short-lived waste is frequently locked in concrete or bitumen before shallow burial. Longer-lived waste is sent to deeper depositories.

HLW is produced in the core of a nuclear reactor, it consists of uranium, fission products, and transuranic elements—it is both highly radioactive and thermally hot. It accounts for 95% of the radioactivity produced in a nuclear power station. Currently the amount of HLW is increasing by approximately 12,000 tonnes per year; the bulk of which is uranium. This would be greatly reduced with the advent of Generation III+ reactors.

The current favoured options for disposition of nuclear waste are:

1. plutonium, minor actinides, and some reprocessed uranium – nuclear fission in a fast reactor;
2. long-lived fission products, such as technium-99 and iodine-129 – transmutation by neutron bombardment;
3. bulky low-level waste – set in concrete or bitumen and shallow burial;
4. medium-lived radionuclides, such as caesium-137 and strontium-90 – secure short-term storage;
5. long-lived fission products and activation products – deep geological secure storage.

The process can be made more efficient by the separation of products with specific treatment relative to their nature. For example, simply heating the waste to a temperature of 1000°C drives off the caesium, which is responsible for half the heat generated during the first hundred years of cooling; this can be safely captured and sequestered. Its removal allows simpler handling of the thermally cooler residue.

The world's largest user of nuclear power, the USA, currently does not reprocess its fuel and hence produces much larger quantities of radioactive waste. In the absence of an agreed federal policy on waste disposal, this waste has accumulated for over 60 years. The US Department of Energy has declared that there are millions of gallons of radioactive waste, thousands of tonnes of spent nuclear fuel, and huge quantities of contaminated soil and water—much of it associated with earlier military activity including weapons testing.

In 1982, the US Nuclear Waste Policy Act proposed the development of a national site for spent nuclear fuel and high-level radioactive waste. In 1987, that site it was decided should be at Yucca Mountain in Nevada on federal land adjacent to the Nevada nuclear weapons test site. In 2001, the US Environmental Protection Agency had set radiation standards for a 10,000 year storage period. These standards were challenged by various interest groups. In 2004, the Court of Appeals accepted all the EPA's standards bar one. The court found that the EPA's 10,000 years was inconsistent with the National Academy's recommendation that the storage lifetime should be increased to a million years. In 2008, the Nuclear Regulatory Commission completed its checking of the DOE's construction and operation license application. The NRC has a statutory requirement to complete its safety analysis and hold public hearings by 2012. Thus the earliest date that construction could start is 2013 with an estimated completion date of 2020.

The Yucca Mountain proposal is for a deep geological repository. This involves a main tunnel 7.6m wide and 8km long. The tunnel is U-shaped with a northerly entrance portal and a southern exit. It is to have a number of smaller side tunnels and several large caverns. The site is licensed to take 63,000 tonnes of heavy metals in spent fuel. It is estimated that the existing nuclear power plants will generate this amount of waste by 2014 unless the USA begins to reprocess its nuclear fuel.

The geological survey of the site cost $9 billion and the drilling machine, similar to that used to create the Channel rail tunnel between England and France, cost $13 million.

The Yucca Mountain project was thrown into confusion by President Obama's declaration that this site was no longer an option and his intention to bring forward a new act to replace those of 1982 and 1987. In 2009, Congress defied this intent by voting overwhelmingly to continue funding the project through 2010; indicating the strong desire to see progress on radioactive waste management.

Since 1999, the Department of Energy has operated a Waste Isolation Pilot Plant for disposal of the Defense Department's transuranic waste. Situated in the Chijhuahuan Desert, New Mexico, the waste is stored in caverns carved in salt deposits more than a kilometre underground. The salt fields have been structurally stable for a quarter of a million years. The waste that can be stored in this way is limited to LLW and ILW because HLW is usually thermally hot and this would attract water which, as concentrated brine, would corrode the steel storage containers in less than the planned lifetime of 10,000 years.

The principal reprocessors of radioactive waste are France and the UK. Both countries receive material from other countries and after reprocessing return the raffinate to the country of origin for final disposition.

There are several methods for extracting specific components of the waste. Additional chemical reactants added following the PUREX treatment can remove the actinides, greatly reducing the alpha radioactivity of the raffinate. Volatilization is used to extract volatile components. Some components become volatile upon heating the raffinate and some by adding chemicals which form volatile compounds.

The securing of high-level waste in concrete or bitumen is inappropriate because the long-term integrity of the material cannot be assured. At the UK's Sellafield site, these materials are locked into glass by vitrification, much in the same way that lead is added to glass to produce crystal. The product is a borosilicate glass, similar to Pyrex. In Russia, a similar process yields a phosphate glass. These glasses are extremely resistant to water. The glass is then placed in welded stainless steel containers for long-term disposition. The UK Committee on Radioactive Waste Management (CoRWM) reported in 2006 in favour of a staged approach to the management of waste including the decommissioning of reactors. The decommissioning process is slow and occurs in stages. The UK Nuclear Decommissioning Agency plans on an average of 50 years per reactor. The long timeframe makes reliable cost estimates extremely difficult. The process is expensive but much less than the value of the electricity generated over the lifetime of the reactor (see Chapter 6).

A typical programme would be to leave the reactor *in situ* for three to five years following cessation of power generation. This allows the core to thermally cool and lose a lot of short-lived, high radioactivity. The site material is then separated into LLW, ILW, and HLW. The LLW is then disposed of as described earlier. The ILW is removed to an interim secure store where it may be held for up to 50 years or so. The interim store is to be managed for heat control and the removal of specific components. The HLW is vitrified and prepared for long-term deep geological disposal.

Arguments continue about the length of time the long-term storage should be planned for (10,000 to 1,000,000 years) and the level of radioactivity that can be tolerated at each stage. In general, conditions imposed in Europe are currently more stringent than those in the USA.

Chapter 5
Nuclear safety

All human activity has associated risk. Before the Industrial Revolution, apart from war, these risks were largely personal. After industrialization, the ramifications of risk are much more widely spread.

The assessment of risk can be based on the analysis of experiential statistics or on a personal subjective basis which may, or may not, have a basis in fact. A large element in subjective risk assessment is fear of the unknown. Familiar activities will be assessed as having a lower risk associated with them than unfamiliar ones. The hazard factor can be high, but the risk can be mitigated by safety precautions.

Nuclear power deals with hazardous materials. It has its origins in a weapons programme that produced the most horrendous capacity for mass destruction. A major source of danger is radioactivity which, for most people, is an unfamiliar and mysterious phenomenon. Bad news headlines sell newspapers and the media frequently report nuclear incidents in more alarmist language than other more frequent industrial accidents would attract. The horrors of Chernobyl would seem to confirm people's worst fears.

The aim of our discussion of 'nuclear safety' is to explain the risks and to present them in perspective.

The hazards of nuclear power begin with the mining of uranium. As with all mining, there are the usual hazards augmented by concerns about the radioactive nature of uranium. In nature, uranium occurs in relatively low concentrations, and thus mining it requires very large volumes of rock to be processed. Most uranium is from open pit mines. Following the exposure of the ore by drilling and blasting, it is extracted via loaders and trucks. In the form of solid ore, the low concentration ensures that there is no direct radiation hazard. However, following extraction and drilling, dust can be inhaled, and its accumulation in the body presents a serious health hazard. Workers should wear face masks at all times. The fact that they spend much of their time in enclosed cabins and vast quantities of water are used to suppress the airborne dust levels greatly reduces the exposure to radiation.

Deep-mined uranium presents a more serious risk. Uranium is an alpha particle emitter and its radioactive decay chain produces radon-222. Radon is an inert radioactive gas with a half-life of 3.8 days. As a gas, radon can be inhaled and its alpha decay is known to be carcinogenic. The problem does not exist in open-pit mining where dangerous accumulations of radon do not occur due to natural ventilation. Underground, in deep pits, accumulations can be hazardous unless adequate ventilation is installed.

Early miners of uranium were reported to have developed small cell carcinoma and some American Navajos, who mined uranium in the south-west USA, were awarded compensation in 1990 under the US Radiation Exposure Compensation Act. In 2008, Areva attracted criticism for not alerting mineworkers at its uranium mine in northern Niger to the health hazards, and air, soil, and water were found to be contaminated.

Nearly 45,000 tonnes of uranium are mined annually. More than half comes from the three largest producers, Canada, Kazakhstan, and Australia. Significant amounts are also obtained from Namibia, Russia, Niger, Uzbekistan, and the USA.

Before the Second World War, scientists researching the new nuclear sciences were not fully aware of the dangers that they faced and many were exposed to harmful amounts of radiation. During the Second World War, women who painted numerals on aircraft instruments with radioactive paint so that could be seen in the dark frequently licked their brushes in order to get a fine point. Several of these workers developed mouth cancers. After the war, the principal users of nuclear material were the military. Clearly not all the personnel understood the dangers inherent in the use of this material. Warheads of enriched uranium or plutonium were machined, as are other metals, on lathes and drills that leave fine particles in the oil that was used to cool the metal during this operation. The standard procedure for getting rid of machine oils was to simply burn it. In the case of nuclear filings, this resulted in plumes of radioactive dust that were carried from the states of Arizona and New Mexico on wind streams from the south to be deposited as far north as Utah. This resulted in cancers in sheep and humans. In a number of cases, ordinary soldiers were taken to witness nuclear weapons tests without being given the appropriate radiation protection, resulting in a number of claims for compensation. As these incidents were reported, they increased the public's concern about nuclear safety.

The most serious reactor failure in the UK occurred at the Windscale site in the north of England. When the Second World War ended, the USA closed its nuclear weapons programme to all other countries, including its partners in the Manhattan Project. The UK embarked on its own nuclear weapons programme and built two graphite piles at Windscale in order to produce plutonium. At the time, relatively little was known about the consequences of exposing graphite to intense neutron fluxes apart from the fact that the graphite was degraded (see Chapter 3).

The Hungarian physicist Eugene Wigner had studied the effects of neutron degradation of graphite and shown that it could lead to the release of energy, causing hot spots in the graphite. To stop the graphite degradation threatening the structural stability of the core and to prevent a build-up of Wigner heating, the graphite was periodically heated to 250°C to promote annealing of the neutron-induced cracks. The fission process generated considerable heat and the pile was cooled by a high flow rate of cold air. The cooling system was not designed to deal with the heat required for the annealing process.

In October 1957, Windscale Pile 1 began to overheat. The fuel for the reactor was metallic uranium which, unlike the uranium oxide used in modern reactors, combusts at high temperature. To cool the reactor down, the operators turned up the flow of cooling air, not realizing the some of the graphite and uranium was already alight. The fresh air turned the core of the pile into a blast furnace. When it was realized what was happening, the air was turned off and replaced by carbon dioxide, but the supply was insufficient to check the fire. Finally, the fire was extinguished by water. The use of water was problematic; if the temperature was too high, the heat could cause the water to dissociate into oxygen and hydrogen, and a build-up of hydrogen could have led to an explosion that would have spread components of the pile widely over the surrounding countryside. This did not occur. Nevertheless, there was a significant release of radioactivity into the surrounding area, and it has been estimated that more than 200 people may have been affected.

The civil nuclear power programme gives rise to concerns about hazards to the general public, including the possibility of an explosive incident, either from plant failure, terrorist attack, or other external event at a power station, resulting in the widespread distribution of highly radioactive material through the air, water, and soil, the theft of radioactive material allowing terrorists the possibility to make an atomic bomb or a dirty bomb (i.e. a conventional bomb clad in a blanket of highly radioactive material that is scattered in the blast), and leaks of radioactive material from a nuclear installation (a power plant, reprocessing facility, or a waste depository).

First, we can absolutely rule out the possibility of a nuclear explosion; nuclear power stations do not have weapons-grade fuel on site.

The designs of nuclear installations are required to be passed by national nuclear licensing agencies. These include strict safety and security features. The international standard for the integrity of a nuclear power plant is that it would withstand the crash of a Boeing 747 Jumbo Jet without the release of hazardous radiation beyond the site boundary.

While there are clearly examples of rogue state removal of reactor core material from which to produce weapon-grade material, this is not a trivial process. It does, however, give rise to concerns about the proliferation of nuclear weapons. To meet these concerns, the Nuclear Non-Proliferation Treaty was opened for signature in 1968. Currently, 189 countries are signatories to the treaty. Of them, only the five permanent members of the UN Security Council (the USA, UK, France, Russia, and CPR) have nuclear weapons. North Korea initially signed the treaty, but withdrew in 2003 when it exploded its own nuclear bomb. Only three other recognized sovereign states are not signatories – India, Pakistan, and Israel. India and Pakistan have tested nuclear weapons, and Israel has a policy of secrecy regarding its nuclear weapons programme. The safety of the world with regard to the threat of nuclear weapons thus rests with the UN Security Council and the influence that it can bring to bear on member states. The test case for this is the UN discussions with Iran over its uranium-enrichment programme.

It is unrealistic to believe that a terrorist cell could attack a nuclear power station and remove material from an operating reactor core. There must be a fear that a terrorist organization could steal, or buy, from a rogue operator weapon-grade material from the stockpiles of plutonium from Cold War weapons that have been decommissioned. The safe remedy is the rapid utilization of these stockpiles in the manufacture of reactor fuels or permanent disposition in secure storage.

In 2003, the IAEA (International Atomic Energy Agency) reported that there were no detrimental health effects or adverse environmental impacts from short- or long-term exposure to depleted uranium weapons material.

In July 2007, Tokyo Electric Power Company's Kashiwazaki-Kariwa nuclear power station was hit by a earthquake registering 6.8 on the Richter scale. Operators were first alerted by a fire in an electrical transformer room. At the time, four of the plant's seven reactors were operating, and they all automatically shut down. The spent fuel from the reactors was stored in water pools. Twelve hours after the shock, operators were alerted to contaminated water leaking from the storage pools through cracks. In all, 315 gallons of contaminated water entered the ocean. The radiation leakage was estimated at one-billionth of Japan's legal limit. The plant was closed for nearly two years while safety checks were carried out.

In August 2009, the Chubu Electric Power Company's Hamaoka plant near Nagoya was hit by a 6.5-magnitude earthquake. Two operating reactors shut down automatically and a third was closed for routine maintenance. It is reported that there was a temporary increase of radioactivity in one of the reactors but no radioactive leakage. Post-shock investigations revealed 39 problems, including malfunctions of neutron monitor and auxiliary transformer operations.

In both Japanese incidents, the automatic shut-down of the reactors was as a result of fail-safe engineering. These require the reactors to be engineered according to 'fail-safe' principles.

That is, if the reactor begins to operate outside its design parameters for whatever reason – human error or faulty equipment – it closes itself down. The two most widely reported incidents in the history of nuclear power had demonstrated the essential need for this feature to be implemented in all reactor designs.

On 29 March 1979, a reactor at Three Mile Island, Pennsylvania, went out of control. For the next week, it was the headline news in every paper across the world. Again, speculation was rife. Unless the reactor was brought under control, the pressure vessel would rupture and a plume of radioactive gas would roll across the state, resulting in thousands of deaths. It did not happen. A popular bumper sticker of the time read 'more people were killed at Chappaquiddick [Chappaquiddick is where an aide to Senator Edward Kennedy was drowned] than at Three Mile Island'. In this regard, the fail-safe engineering of Three Mile Island reactor behaved according to its specification.

The importance of the 'fail-safe' principle was highlighted seven years later at Chernobyl in the Ukraine, then part of the USSR. The totalitarian Soviet regime did not encourage a questioning of those in authority. Where in the West protest groups are often an irritation to the establishment, but can freely challenge orthodoxy and influence policies, protests in the USSR could lead to banishment to the Gulags. At Chernobyl, the authority of the reactor operators was not questioned when they began an illegal reactor experiment in May 1986. The reactor began to run out of control and the lack of fail-safe engineering resulted in all the worst fears of Three Mile Island being realized. A build-up of hydrogen, a cause of concern during the Windscale fire, led to an explosion that blew the plant apart, with a large release of radioactivity. Two people died in the initial blast and 29 died from the immediate radiation burst. A further 200 people were treated for radiation burns and sickness. There was a great fear that exposure to the radiation would harm a much larger population for decades to come. Now, 20 years later, the impact appears to be less severe than originally feared, with fewer than 60 deaths directly attributable to the incident. The public harm appears to be limited to fewer than 2,000 cases of adolescent thyroid cancer, which is normally curable. It is true that the land once used for agriculture remains unsafe for food production and that this is a long-term consequence of the incident.

In both cases, the accidents were due to human failure. At Three Mile Island there were no injuries and the only losers were the reactor operators, for whom it was a financial disaster. We now know that the cause of the incident was the consequence of greed and the US tax regulations. If the plant operators could commission the plant by 31 December, they would get a full year's tax allowance. Thus the plant was rushed into service before every component had been fully checked and the staff fully trained. The incident was triggered by a combination of a faulty component and an inexperienced human response. The impact on the public perception of the dangers of nuclear power was increased by the Jane Fonda film *The China Syndrome*, a fictional version of a Three Mile Island type of event, released just a few days before its actual occurrence. Nearly 30 years were to elapse before proposals for new nuclear plants in the USA were to appear. At Chernobyl, an unauthorized experiment at a commercial nuclear station that was not fail-safe engineered resulted in disaster.

Why should we draw any comfort from such an explanation we have given for the incident at Chernobyl? Could a similar event not occur in the future with devastating consequences? It has to be stressed that the accident was completely avoidable if fail-safe engineering had been employed; the Chernobyl reactor would never have been licensed in North America or Western Europe. A nuclear plant in Finland originally of the Chernobyl design has been retro-fitted with fail-safe technology and continues to operate safely. The lessons to draw from these two incidents are that humans are fallible and that reactor designs should only receive licences for construction if they are adequately fail-safe designed. Indeed, they give great weight to the argument that national licensing of nuclear plant is an insufficient international safeguard. Since failure of a nuclear plant has the potential to produce damaging effects beyond the borders of the country in which it is operated, the licensing of the design should be subject to IAEA approval.

However, no matter how careful the planning or the regulatory framework, or how excellent the construction, not all risks can be eliminated. This was graphically illustrated when the east coast of Japan was struck by an earthquake and ensuing tsunami on 12 March 2011. The final total death toll was in the order of tens of thousands. Hundreds of thousands of people saw their homes and belongings totally destroyed, along with hospitals and schools. Large sections of the national infrastructure of road and rail systems and water and electricity supply suffered considerable disruption. There were major explosions and fires at gas and oil storage facilities. The final economic cost of this disaster is likely to top a trillion US dollars. Japan is one of the world's largest economies, and the economic impact quickly spread across the globe. In the midst of all this destruction and chaos, concerns naturally arose about the impact on Japan's nuclear power plants. We have already noted the earlier impact of earthquakes on the Kashiwazaki-Kariwa plant near Tokyo in 2007 and the Hamaoka plant near Nagoya in 2009. Both these earthquakes measured less than 7.0 on the Richter scale. The March 2011 quake measured 9.0 on the Richter scale and triggered a devastating tsunami.

There were 17 nuclear power sites in Japan operating 55 nuclear reactors, most of which were BWRs. Concerns quickly concentrated on the Fukushima-Daiichi plant north of Tokyo. The plant operated six BWRs built in the 1970s, and it was obvious that the cooling systems had been seriously compromised. As temperatures in the reactor cores rose, a number of explosions shook the plant, but the containment vessels remained intact. In addition to the fuel rods overheating, the highly radioactive spent fuel rods, kept in storage pools, started to heat as the pools lost water. Externally, water and electricity supplies to the plant had been cut. The authorities struggled to restore power and water supplies to the plant.

There was some radioactive leakage from Fukushima detected in Tokyo, and drinking water was briefly contaminated with radioactive iodine. Raised radioactivity levels were also detected in spinach crops and milk from farms close to the plant. The authorities moved rapidly to remove these products from the market, though experts concluded that these had posed negligible risk to human health.

An emergency 30-km exclusion zone was established around the site and, as a precaution, the region was evacuated, adding further to the difficulties arising from the nearly half-million refugees already rendered homeless by the earthquake and tsunami.

As with previous nuclear incidents, a full international enquiry was initiated and lessons will be learned. But the enquiry may take several years to complete. (The site of the Windscale fire is still being monitored 50 years after the event!)

Some sections of the media made comparisons with Chernobyl. It is relevant to compare the worst-case scenario at Fukushima with Chernobyl. The worst-case scenario was that a hydrogen-based explosion sufficiently powerful to destroy the already weakened containment vessel occurred. Because of the different construction of the two reactors, the blast would have been less strong than that at Chernobyl. The blast consequences for personnel in the reactor at the time of such an explosion were likely to be similar. However, the developments at Fukushima were fully monitored, and it may have been possible to evacuate all personnel before an explosion. While the long-term impact on personnel in the reactor buildings will take years to assess, and some fire-fighters were exposed to intense radiation levels, they were wearing the best radiation-protection garments available. The evacuation of the region around the site would have greatly reduced the radiation hazards to the local population. The Chernobyl explosion released a plume of radioactivity that was driven north and west by prevailing winds across the highly populated European continent. The plume was detected in Scandinavia. As a precaution, meat and dairy produce from contaminated grazing land was taken off the market. There has been no suggestion that anyone outside the Ukraine suffered harm from the radiation plume. A plume from a Fukushima explosion would naturally be driven eastwards over the Pacific Ocean and, although it might have been eventually detected on the west coast of North America, it would not have been harmful to humans.

Unlike Three Mile Island and Chernobyl, there is no suggestion of human failure in the operation of the Fukushima plant. At Three Mile Island, not all components of the reactor had been fully tested before operations began, and Chernobyl was not constructed to current fail-safe engineering standards and was the subject of an illegal experiment. At Fukushima, the design was to current safety standards, taking into account the possibility of a severe earthquake; what had not been allowed for was the simultaneous tsunami strike. It is recognized that human activity in regions of geological instability carries additional risk, and Fukushima serves as a timely reminder that the forces of Nature can still upset the best-laid plans of humankind.

While Fukushima dominated the headlines, the nuclear component of the tragedy that followed in the wake of the earthquake and tsunami in Japan represented a very tiny fraction of the cost, in both financial and human terms, of the disaster.

Nothing short of a human disaster can categorize the incident at Chernobyl. However, it should be seen in relation to other large-scale industrial activity. For example, in December 1984 the Union

Carbide pesticide plant in the city of Bhopal in the Madhya Pradesh state of India released 42 tonnes of toxic gas. More than half a million people were exposed. Approximately 9,000 people died within the first 72 hours, and it is estimated that 25,000 have died since then from gas-related illness.

There have been many reports of minor incidents at nuclear installations, but none involving the loss of life associated with the nuclear aspect of the plants. At each stage, as the technology advanced, the limits on reportable incidents have been lowered. In the UK, most of these incidents have related to waste storage leaks at the Sellafield and Dounreay sites. Here, waste has been stored in temporary facilities awaiting political decisions on its final disposal. It seems incredible that it took 50 years from the opening of the first reactor to the commissioning by the UK government of the Committee on Radioactive Waste Management and its report.

One of the most alarming incidents was the report of cancer clusters around nuclear sites in the UK, most seriously at Sellafield. The cancer was of an unusual form in that it was among young children. It was proposed that the cause was due to radiation received by fathers working at the nuclear plant before the conception of their children. The reports had a devastating effect on many families. Detailed medical studies have revealed that the cancer clusters had nothing whatever to do with the nuclear nature of the plants but rather is an epidemiological feature of large-scale construction projects in rural areas where there are relationships between the incoming project workers and the local population. While the initial reports were world news, the more mundane explanation has received little publicity.

The effects of radiation are short-ranged and thus can only become more widespread if it is water soluble or airborne. These possibilities carry the additional hazard that the radioactive material can be ingested either from breathing contaminated air or drinking contaminated water. This hazard is amplified in the case of radioactive isotopes that chemically mimic elements that the body relies on for specific functions. Thus iodine is naturally concentrated in the thyroid gland. Radioactive isotopes of iodine are a by-product of fission and do exist in reactor cores, and, as we saw, the bulk of the radiation patients at Chernobyl were treated for thyroid cancers. A common precautionary measure, if there is thought to be the possibility of leak of radioactive material, is to give iodine tablets in advance to saturate the thyroid gland so that it does not accumulate the radioactive isotopes.

Another common nuclear fission product is radioactive strontium. Strontium is chemically similar to calcium, which the body concentrates in the bones. The bones are where blood is created, and thus dairy animals ingesting strontium-contaminated grass may produce milk, a common source of calcium for humans, which carries the risk of producing leukaemia. Indeed, the milk and meat of all grazing animals pose a potential hazard if the pasture has been contaminated by radioactive fallout. The prime concerns in decommissioning a nuclear plant and the safe storage of radioactive waste are to prevent radioactive dust entering the atmosphere and the prevention of radioactive contamination of water supplies.

Throughout discussions of these contentious issues, there are two topics that polarize the debate: is there a safe level of radioactivity that can be tolerated, and how long must the lifetime of the secure store be?

Radioactive strengths are measured in becquerels ($1 Bq = 1$ disintegration per second). In dealing with radioactive materials, the curie ($1 Ci = 3.7 \times 10^{10} Bq$), is still sometimes used. The biological impact of radioactivity depends not just on the number of radioactive emissions but also on their type (alpha, beta, gamma, neutrons) and energy; it also depends on the region of the body receiving the radiation; for this reason, radiotherapy is carefully focused on the source of cancer tumours. For levels of radiation falling within normal recommended safety limits the severity of the biological impact is measured in units of sieverts. This is given in the case of whole-body exposure by the equivalent dose, which is the energy deposited in joule/kg multiplied by a radiation weighting factor between 1 and 20 depending on the ability of the particular type of radiation to cause damage.

Nuclear radiation is a part of our natural environment. It comes to us from the Sun and the stars and from the rocks beneath our feet. Indeed, it is an essential component in the evolution of plants and animals. Natural radiation can randomly cause minor biological mutations. Most of these are short-lived and not passed down through the generations; some may produce cancers, and some produce genetic modifications that are passed to future generations. Successful mutations provide benefit to the individual and natural selection sees these mutations passed on.

New York may be considered to be a hazardous city, but few would consider radiation from the granite of Grand Central Station, which exceeds the current safety limit imposed on the operators of nuclear power stations at the edge of their sites, to be a significant danger. Mountaineering is a hazardous activity, but the increase in radiation exposure with altitude is not a danger often considered.

Most people love the sunshine. At one level, it provides the heat and light that make our lives possible. We are all aware that overexposure presents a hazard in the form of sunburn or, *in extremis*, melanoma (skin cancer). However, we are also aware that protection can be effectively provided by a simple barrier cream. The beaches of the world attest to the confidence that most people have handling the dangers of solar radiation which arises from the nuclear activity in our Sun.

Cosmic rays interact with our atmosphere to produce a radioactive form of carbon (carbon-14, or radio carbon). This radioactive carbon is completely mixed with stable carbon-12 in our atmosphere and is taken up by plants during normal photosynthesis. Thus, all our food naturally contains radioactive material, as does the air we breathe. This is not a cause for concern. It may be that cosmic rays and their radioactive products do cause mutation in an individual cell, a process that has allowed evolution to occur, but the risk of serious health implications is negligible.

We have no choice but to live with this natural radiation background. Our atmosphere protects us from most of the solar and cosmic radiation, and sunblockers and hats from the rest. Buildings in granite areas and built of granite may now be checked for ventilation to ensure that there is no accumulation of radon gas.

In setting 'safe' levels of radiation for human activity, this variable natural background is a useful yardstick. The worldwide average annual natural human dose from the natural background is 2.5mSv (2.5×10^{-3}Sv), and varies from around 1mSv to a high of over 200mSv with location. This is to be compared with the human contribution from all nuclear activity (both military and civil) and other non-nuclear industries of 5μSv (5×10^{-6}Sv).

The World Health Organization has declared that the largest human activity contributor to global radiation is not the nuclear industry but fly-ash from burning coal, which it estimates causes 350 deaths from radiation per year globally. The largest nuclear contribution comes from early above-ground testing of nuclear weapons. This contribution peaked at 0.15mSv in 1963 and is declining at a rate of 0.005mSv per year.

With regard to the question 'what should the safe radiation limit for the general public be?', there are those who hold that if there is any detectable source of radiation then this is unacceptable. As our capacity to measure lower and lower levels of radiation has increased, safety levels have steadily been lowered. National limits vary, but are currently set at around 1 mSv/year. In this debate there are no absolutes. The impact of radiation, both natural and produced by humans, is statistical in nature. It should be recalled that radioactive material never completely stops radiating. When we talk about the half-life of a source of radioactivity, we mean the time during which the level of radiation from that source halves. After several such periods, the level has fallen to such an extent that we consider it 'safe'. All human activity carries a risk; the question is, to what extent should that risk be mitigated?

With regard to secure, safe radioactive waste disposal, the question is 'how long must the HLW store be secure?' Initially, the USA proposed 10,000 years. Currently, more stringent limits of a million years are being suggested. Again, there are no absolutes. A personal view is that looking back over human development in the past 10,000 years, I can have little concept of the world 10,000 years in the future, far less a million years.

The nuclear industry is rightly the most stringently controlled industry in the world. Incidents involving nuclear plant are widely reported. Many incidents at nuclear plants are due to engineering failures unconnected with the nuclear aspects of the site. A cull of all reported civil nuclear incidents worldwide reveals that, excluding the Chernobyl disaster, in the 1950s one nuclear plant worker in Yugoslavia developed leukaemia, in the 1980s two plant workers were killed in Argentina, and in 1999 two plant workers died in Japan. Despite the fact that the records may not be complete, principally due to the secrecy surrounding such incidents in the former USSR, the safety record of nuclear power stands favourable comparison with any other global industry of a similar scale.

Chapter 6
The cost of nuclear power

The costing of nuclear power is notoriously controversial.
Opponents point to the past large investments made in nuclear
research and would like to factor this into the cost. There are
always arguments about whether or not decommissioning costs
and waste-management costs have been properly accounted for.
When it comes to comparative costing, additional arguments arise
because there is no universally agreed basis for a common
procedure for different sources of power. The situation is further
complicated by differential tax regimes applied in different
countries to different power technologies; some are taxed heavily
while others are given government grants. One of the growing
differentiators is 'green' taxes designed to reduce dependence on
fossil fuels and the greenhouse gases that they emit. Nuclear power
is a non-carbon gas emitter but seldom enjoys the tax breaks
provided for the so-called renewable sources of power.

In comparing costs between different countries, we have to be
aware of local labour and material costs. In comparing costs across
different generating technologies, we again must be conscious of
local available alternative energy sources, for example geothermal
energy in Iceland, hydroelectricity in Norway, coal in Poland,
oil and gas in the Middle East, and so on. Thus the decision on
which electricity source is most economical will vary from
country to country and there is no magic formula for all.

As with all industrial processes, there can be economies of scale. In the USA, and particularly in the UK, these economies of scale were never fully realized. In the UK, while several Magnox and AGR reactors were built, no two were of exactly the same design, resulting in no economies in construction costs, component manufacture, or staff training programmes. The issue is compounded by the high cost of licensing new designs. France led the way by adopting a single reactor design and introducing a 'type' licensing scheme similar to that used for new aircraft. Public debate was thereafter limited to local site issues. As a result, in 2006 the cost of generating nuclear energy in France was two-thirds of the cost in the UK.

With these health warnings we shall proceed.

Let us begin with the construction costs. These are the largest single factor in the cost of nuclear electricity generation. These capital costs have to be paid before the power station has produced any income from the generation of the electricity it does produce. The cost has to be defrayed over the lifetime of the reactor, and the largest element in this cost is the cost of providing the required capital; just like the mortgage interest paid on buying a family home.

Decommissioning costs can be estimated at the construction phase but may not accurately reflect the actual cost 40 or 50 years later. Similar uncertainties surround the estimates of waste management. However, these uncertainties are extremely small compared with the capital costs.

In both the UK and the USA, there were considerable over-runs on projected capital costs in the 1960s to 1980s. There were three principal reasons for this:

1. Following incidents like that at Three Mile Island, regulatory authorities, upon reviewing the safety of nuclear plant design,

ordered changes that were costly and time-consuming to implement and added significantly to the capital cost of construction.

2. The bureaucracy between getting a licence to build and a licence to operate new nuclear plant, often involving extensive public participation, delayed construction for many years and added to the cost. This was a major factor in the cost of the UK's first PWR at Sizewell and an even bigger issue at the Shoreham Plant on Long Island which, after an expenditure of $5 billion, was refused an operating licence.

3. Non-uniform designs meant that in both countries there were no economies of scale. Equipment had to be custom-built for each plant and each design had to obtain separate licensing approval.

As we have seen in France, the Regulatory Commission agreed a standard design for all plants and used a safety engineering process similar to that used for licensing aircraft. Public debate was thereafter restricted to local site issues. Economies of scale were achieved. Construction was on time and there was little in the way of over-run on capital costs.

The claim that Generation III reactors can be built faster than earlier PWRs is reflected in the claims for construction times of 3.5 years for the ACR from order to operation, 5 years for the AP1000, 4 years for the EPR (from first concrete to operation, add site preparation time), and 45 months for the ESBWR, compared to earlier reactors which frequently took more than 10 years from order to operation.

The first Japanese ABWR was completed close to both capital cost and on time, but the second two are subject to unexpected delays. The first EPR being built in Finland has hit unexpected delays and will be two years overdue.

In the USA, the NRC has streamlined its licensing procedures so that a single construction and operating licence will be awarded, and the new plants will have final design approval before construction begins. In the event that alterations are required following the start of construction, the Federal Agencies will share the costs of over-run with the constructors. Both Japan and the USA have adopted the French safety engineering reactor-type licensing. The UK has proposed similar changes to its licensing procedures.

Because of the history of the nuclear power industry, the financial sector sees it as a risky investment and demands a premium rate for the capital required. In 2005, a joint study by the OECD (Organisation for Economic Co-operation and Development) and the NEA (Nuclear Energy Agency), assuming a 10% discount (interest) rate, concluded that capital costs represented 70% of the cost of generating the electricity and that the discount rate was the most important parameter in determining the cost of electricity over the lifetime of the reactor. To give comfort to the investors in new nuclear plant, the US Congress has voted on a subsidy, comparable to that offered to the wind power industry, for the first three years of operation.

The Generation III+ reactors are claimed to be half the size and capable of being built in much shorter times than the traditional PWRs. The 2008 contracted capital cost of building new plants containing two AP1000 reactors in the USA is around $10–$14 billion, depending on the degree of inclusion of costs for land acquisition, site preparation, cooling towers, finance, licensing and regulatory fees, initial fuel load, insurance and taxes, escalation, and contingencies.

The Chinese nuclear power industry has won contracts to build new plant to its own design at a capital cost approximately one-third that of the AP1000s in the USA. The cost of labour is much lower and the financial arrangements and licensing are not yet clear. If these plants are constructed within budget, they will present a major challenge to the import of Western reactors.

A conservative 2006 UK study suggested that the capital costs of a new AP1000 or EPR would contribute 1.7p/kWh to the cost of electricity over the lifetime of the reactor.

By comparison, the cost of fuel is much more straightforward. The fuel for PWRs is in the form of enriched uranium dioxide.

In the USA at the beginning of 2007, to produce 1kg of fuel, required 9kg of yellow cake (triuranium octooxide, the raw basic fuel material), costing $472, conversion to pure uranium cost $90, enrichment cost $985, fuel fabrication cost $240. Thus the total cost of the fuel was $1787/kg. This would contribute 0.5c/kWh to the cost of electricity.

In 2008, Areva gave a breakdown of EPR costs as being 17% from fuel, of which 51% for raw material, 3% conversion, 32% enrichment, and 14% fuel fabrication. The higher cost of enrichment is presumably due to the higher level of enrichment required for the advanced PWRs.

Unlike fossil fuels, the cost of nuclear fuel has been falling. In Spain, the cost of fuel was reduced by 40% between 1995 and 2001 by boosting enrichment to achieve greater burn-up.

The UK 2006 study estimated that fuel, including spent fuel reprocessing, contributed 0.46p/kWh to the cost of electricity.

Because the raw fuel is such a small fraction of the cost of nuclear power generation, the cost of electricity is not very sensitive to the cost of uranium, unlike the fossil fuels, for which fuel can represent up to 70% of the cost.

Operating costs for nuclear plants have fallen dramatically as the French practice of standardization of design has spread. In the West, there are now basically only Areva, Westinghouse, General Electric, and AECL offering reactor designs. This has resulted in economies of scale for common component costs, shared experience and training programmes. Current USA and European operating costs are consistent (2006) at approximately 1p/kWh.

There is considerable experience of decommissioning of nuclear plants. In the USA, the cost of decommissioning a power plant is approximately $350 million. The funds to pay for this are accumulating from the charges made from the sale of electricity. In France and Sweden, decommissioning costs are estimated to be 10–15% of construction costs and are included in the price charged for electricity. Similarly, in Finland, the power plant operators are required to create a decommissioning fund from operating income. The UK has by far the highest estimates for decommissioning which are set at £1 billion per reactor. This exceptionally high figure is in part due to the much larger reactor core associated with graphite moderated piles. This was allowed for in the retail cost of electricity but as a nationalized industry, no sinking fund was created to cover the decommissioning costs. This was a problem when the energy industry was privatized and, quite naturally, private companies were unwilling to inherit the decommissioning costs without the accumulated income.

Following decommissioning there is the cost of final waste disposal. In the USA, nuclear plant operators are charged 0.1c/kWh and in Sweden 0.13c/kWh. To date, both countries have used this income to fund research leading to their deep geological waste disposal plans. In France, waste disposal and decommissioning charges are set at 10% of construction costs. The accumulated fund from the sale of electricity is close (2009) to 80 billion euro.

Combining all these costs, we arrive at the cost of electricity generated from nuclear fission.

	%Disc	%Load	Years	P/kWh
MIT	11.5	85	15	4
PRINCETON	11.5	80	15	4
CHICAGO	12.5	85	15	3.5
RAE	7.5	90	32.5	2.8
F ECON M	8.0	90	42.5	2
FINLAND	5.0	90	40	1.7
OECD 1	5.0	90	40	1.6
OECD 2	10.0	90	40	2.8

Table 2. The cost of nuclear power

We present a series of estimates made between 2000 and 2005 in Table 2. The costs of nuclear-generated electricity were estimated by three US university studies (MIT, Princeton, Chicago), three EU studies (UK Royal Academy of Engineering, French Ministry of Economics, Finish Lappeerata University), and two OECD studies.

The higher costs from the US studies compared with the European studies, and the difference between the two OECD studies, are clearly linked to the cost of capital. The first column is the cost of borrowing the capital, the second is the assumed annual operating load as a percentage of capacity, the third is the number of years over which the reactor is operated, and fourth is the cost of the electricity generated in UK in pence per kWh.

In 2005, the OECD/IEA study projected costs of generating electricity in 2010 on the basis of an assumed discount rate of 10%, over a 40-year lifetime at an 85% load factor. In these studies, the cost of coal, gas, and nuclear were compared. The results in Table 3 are in 2003 US cents per kWh.

	Nuclear	Coal	Gas
FINLAND	4.22	4.45	-
FRANCE	3.93	4.42	4.30
GERMANY	4.21	4.09	5.0
SWITZERLAND	4.38	-	4.65
NETHERLANDS	5.37	-	6.26
CZECH REPUBLIC	3.17	3.71	5.46
SLOVAKIA	4.55	5.52	5.83
ROMANIA	4.93	5.15	-
JAPAN	6.86	6.91	6.83
KOREA	3.38	2.71	4.94
USA	4.65	3.65	4.90
CANADA	3.71	4.12	4.34

Table 3. OECD study of relative electricity-generating costs

Across Europe, we see nuclear electricity costs are in the range 3–5c/kWh. In Asia, we see a low cost in Korea but the highest cost of all in Japan. In North America, Canadian electricity is less expensive than in the USA.

Of the ten countries for which a comparison with coal was relevant, nuclear generation was cheaper in six. Of the ten countries where comparison with gas was relevant, nuclear was cheaper in eight.

It is clear that in many countries nuclear-generated electricity is commercially competitive with fossil fuels despite the need to include the cost of capital and all waste disposal and decommissioning (factors that are not normally included for other fuels).

A major European study tried to put a value on external costs, which are defined to be those actually incurred in relation to health and the environment. These estimates did not attempt to include the impact of global warming. On average, external costs (in addition to those in Table 3) across Europe were 0.4c/kWh for hydroelectricity and nuclear power, 2c/kWh for gas, and over 5c/kWh for coal.

As we look to a future focusing on renewable generating costs, we can make comparisons with the cost of electricity generated by wind (on shore and off shore), waves and tides, biomass, geothermal, and solar. The market is heavily distorted by green policies, including taxes increasing on carbon gas emissions and tax breaks to encourage new renewable sources of electricity. Many of these are not geographically transferable. Solar power is greatest between latitudes $40°$ north and $40°$ south. The availability of geothermal energy depends on local geology. Marine power is of little interest to a land-locked country, and hydroelectricity requires large fast-flowing rivers and the land to create vast reservoirs. Wind power is universally available, but the efficiency of electricity generation depends on the specifics of site choice. In all cases, the intermittency of wind requires the building of additional capacity as backup. The cost of these emerging new technologies will fall as the market for them grows and experience leads to greater efficiencies. At the present time, without the market of taxes and grants, electricity generated from renewable sources is generally more expensive than that from nuclear power or fossil fuels.

This leaves the question: if nuclear power is so competitive, why is there not a global rush to build new nuclear power stations? The answer lies in the time taken to recoup investments. Investors in a new gas-fired power station can expect to recover their investment within 15 years. Because of the high capital start-up costs, nuclear power stations yield a slower rate of return, even though over the lifetime of the plant the return may be greater. Investors in nuclear plant require clear government signals that long-term stability in the market is assured, or, as in the USA, a front-end subsidy similar to those available for other 'green' electricity generators. An international agreement on a framework to price carbon emissions would be the single most effective means of encouraging non-carbon-producing technologies, including nuclear power. The practice of 'carbon trading', pioneered in the EU, is not a practical solution to the emission of climate change gases. It simply means that a country that does not exceed its carbon credits can sell the excess to a country that cannot contain its emissions within its allocation of carbon credits. The result is the export of emissions without any net global environment benefit.

Chapter 7
Nuclear fusion power

Whereas the current nuclear reactors release energy by fissioning heavy nuclei to create lighter fragments, and are aided in this by the natural electrical repulsion of the positive charges in the nuclei, a fusion reactor is designed to produce the joining together of extremely light isotopes of hydrogen to form more stable heavier nuclei, which is inhibited by the electrical repulsion between like charges. The prototype of all fusion reactions is the joining together of isotopes of hydrogen to form the extremely stable nucleus of helium-4.

A fission reactor can start from cold. The neutrons that initiate the chain reaction are slow moving. The heat builds up as the chain reaction releases the nuclear binding energy. In the fusion reaction, the hydrogen isotopes have to be moving rapidly enough to overcome their electromagnetic repulsion of each other. To develop a chain reaction the fuel has to be maintained at temperatures exceeding those at the centre of the Sun and at a high density. At these temperatures, the collisions between the atoms are so violent that the electrons are instantly torn from them, creating a gas of free moving negatively charged electrons and positively charged nuclei; this state of matter is called a 'plasma'. The problem is how to contain a plasma; any contact with a material container would lead to its instant melting or loss of heat by the plasma, and hence drop in temperature, such that the fusion reaction ceases.

Such fusion reactions were common in the early universe, but were transitory as the universe expanded and cooled and became unsustainable after a few minutes. Today such reactions are the principal source of energy powering main sequence stars like our Sun. The containment in this case is the gravitational pull of the star in the vacuum of space. There is a minimum size of a star because it must be large enough for its self gravity to compress the matter it contains sufficiently to ignite the fusion process. In the case of our Sun, it is just sufficiently large enough for this to happen. Indeed, the heat put out per unit mass from the Sun is less than the heat put out per unit mass from the human body. This slow burn has allowed the Sun to pour forth its energy for almost 5 billion years; long enough for the solar system to form and life on Earth to develop.

Man has managed to create a fusion reactor on Earth in the form of hydrogen bombs each equivalent to 500 atomic bombs. In the case of a bomb, there is no thought of containment. For the past 50 years, the search has been on for a means of harnessing this enormous source of energy to controlled civil use.

There are two principal contestants in the search for a containment system, confinement by magnetic fields and confinement by inertial bombardment.

The most highly developed confinement system is based on the fact that the plasma comprises freely moving electric charges and hence it may be possible to create electromagnetic fields that will hold it in a vacuum vessel without touching the walls. The first patent relating to nuclear fusion reactions was taken out by the UKAEA (United Kingdom Atomic Energy Authority) in 1946 based on the researches of G. P. Thomson and M. Blackman. The principles contained in the patent, magnetic confinement and radio frequency plasma heating in a toroidal vacuum chamber, are identical to those in the current international programme to develop a prototype fusion reactor.

In 1951, L. Spitzer initiated the US fusion programme at the Princeton Plasma Physics Laboratory. Code-named Project Matterhorn, this programme studied various confinement geometries and especially the 'stellarator' structures (toroids twisted into a figure of eight). These different geometries all led to instability problems in the confinement of the plasma.

In the early 1950s, the Soviet theoreticians I. E. Tamm and A. D. Sakharov refined Thomson and Blackman's analysis by including magnetic coils in addition to the toroidal geometry in a device they named the Tokomak. The Tokomak uses the electric current resulting from the plasma flow to generate a helical magnetic field to produce stability in the plasma. This was so successful that in 1968 the USSR announced the production of the first quasistationary thermonuclear fusion reaction at Novosibirsk.

Various vessel designs and electromagnetic field configurations have been studied but the present leading concept is based on a toroidal (doughnut), or in Russian 'tokomak', design. Currently most magnetic confinement research has concentrated on the Tokomak concept.

In the toroidal fusion reactors, the torus is the vacuum confinement chamber. A constant electric current in the direction of the smaller circles A flows through coils all round the torus and produces a magnetic field in the direction of the circles B. The field is strongest in the centre of the torus. Charged particles moving in a magnetic field spiral along the magnetic lines of force. It is this property that produces the aurora borealis and the aurora australis (northern and southern lights) when cosmic ray charged particles are trapped by the Earth's magnetic field at the poles. Fuel in the form of hydrogen isotopes is introduced into the chamber and heated by external sources such as radio-frequency generators (see below) until it becomes a plasma.

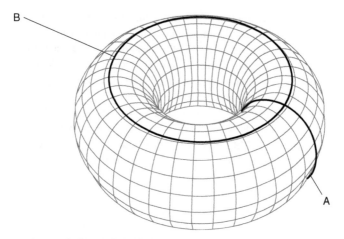

17. A torus is the product of two circles

The charged particles of the plasma then follow helical paths round the torus tied to the magnetic field lines. The stronger the magnetic fields the more tightly are the charged particles tied to the magnetic field. As the plasma circles the torus, it becomes an additional current loop and generates a poloidal magnetic field in the same direction as the toroidal currents. A charged particle moving in a magnetic field experiences a force perpendicular to the field. Thus the poloidal field stabilizes the plasma, keeping it from the vessel walls and circulating round the torus.

The largest of these machines is the Joint European Torus (JET) at Culham near Oxford. This is paving the way for the International Thermonuclear Experimental Reactor (ITER) being constructed in France as a collaboration involving the EU, China, India, Japan, Korea, Pakistan, Russia, and the USA. Construction is expected to take about 10 years, with switch-on scheduled for 2018. It is planned to operate ITER for 20 years, during which time ITER scientists will study plasma conditions expected to pertain to an electricity-generating, fusion power plant. ITER will use 18 toroidal field coils of supercooled niobium-tin to carry a current of 46kA and produce a magnetic field of 11.8T (tesla). It will generate more energy (500MW) for extended periods of time than is required to maintain the plasma temperature and hence will be the first fusion experiment to produce net power. It will also test a number of key technologies and new materials essential for a practical operating power station. While ITER is designed to generate between five and ten times more energy than it consumes, there is no intention to generate electricity at this stage.

ITER is intended to be the final experimental fusion facility on the road to constructing the first prototype commercial fusion power station. It is anticipated that this might be possible by the middle of this century.

The second process is inertial confinement and is a development of the hydrogen bomb principle. In the hydrogen bomb, the hydrogen isotope fuel was compressed very rapidly to high density and temperature by the force of a thermal fission explosion. In a fusion reactor, tiny droplets of fuel are bombarded by high-powered laser beams from all sides such that they are compressed to the point where fusion reactions occur. The intent is to drip feed droplets to produce a continuous output of energy.

The proposal to use laser inertial confinement of the plasma was made by scientists at the US Lawrence Livermore National Laboratory in 1962, shortly after the development of the laser in 1960. However, the early lasers clearly were not powerful enough for the task and it was not until the 1980s, by which time dramatic increases in laser power were possible, that the technique appeared worthy of further consideration. The principal centres for laser confinement research are at the NIF (National Ignition Facility) in California and at the French Laser Megajoule Laboratory.

The leading proposal for the next step in research into inertial confinement is the European HiPER (High-Powered Laser Research Facility) project. Planning began in 2005, received encouragement from the EU in 2007 and the design phase started in 2008 with the intent that construction would begin in 2011–12.

Early attempts at laser confinement proposed to induce a spherical shockwave travelling inwards in the fuel droplet with the intention of driving up the temperature and density in the plasma to achieve fusion ignition. This approach requires extremely powerful lasers, 330MJ of electrical power are required to drive the lasers at NIF in order to achieve an output of 20MJ of fusion power; clearly such a device can never be a practical energy source.

HiPER proposes a different approach called 'fast ignition' in which two laser systems act in tandem. Relatively low powered lasers compress the fusion fuel to a density of about 300g/cm^3 which is only one-third of the density in the NIF shockwave but the lasers only require an input of 200kJ. The compressed fuel is then bombarded by light from a second set of high-powered (76kJ) lasers in a short (10ps) pulse. This pulse interacts with the compressed plasma to produce a shower of highly energetic (3.5MeV) electrons which interact with the plasma to drive it beyond the ignition temperature. While HiPER will require many more lasers than NIF, their costs will be much lower. At NIF, the ratio of fusion power out to laser power in (the fusion gain) is about 5; HiPER expects to achieve fusion gains of between 50 and 100.

The animals in the fusion zoo are hydrogen-1 (protium) consisting of a single proton and comprising 99.985% of natural hydrogen, hydrogen-2 (deuterium) consisting of one proton and one neutron and comprising 0.015% of natural hydrogen and hydrogen-3 (tritium) consisting of one proton and two neutrons, helium-4 consisting of two protons and two neutrons, and helium-3 consisting of two protons and one neutron. Helium-3 is stable but only present in trace quantities on Earth. Tritium is unstable and does not exist in nature outside the centre of stars where they are created and decay during violent nuclear reactions or by collisions between high-energy cosmic rays and the Earth's atmosphere. Tritium has a half-life of about 12 years and so does not accumulate on geological timescales rendering its natural abundance essentially zero. However, a half-life of 12 years is more than long enough for tritium to play a role in a fusion reactor. Unlike fission fuel dependent on uranium-235 and uranium-238, where the small relative mass difference makes the separation of isotopes difficult, the large relative mass differences between the isotopes of hydrogen make their separation straightforward.

The most studied fusion reactions are deuterium on deuterium and deuterium on tritium. Figures 18, 19, and 20 show schematics of the principal reactions involved in a fusion reactor. The velocity of the neutron and the helium nucleus are measures of the energy released in this reaction. The release of 17.6MeV shared between five nucleons in the deuterium–tritium reaction should be compared with the 0.7MeV per nucleon released in a typical fission reaction.

The deuterium–deuterium reaction unfortunately does not lead directly to the formation of helium-4, but to the production of helium-3 plus a neutron or tritium plus a proton in roughly equal amounts. The energy released in the production of helium-3 and tritium is much less than from the production of helium-4, but the tritium produced becomes fuel for the deuterium–tritium fuel cycle.

The reaction can be made self-sustaining by introducing lithium (three protons) into the plasma. Collisions between the neutrons and lithium-6 produce tritium and helium-4, adding to the energy release but also replenishing the tritium. Collisions between the neutrons and lithium-7 result in the production of helium-4, tritium, and a neutron. This is an endothermic reaction, that is, it uses some of the neutron energy but maintains the level of tritium and neutrons in the plasma. While lithium is much less abundant than hydrogen, there is enough lithium ore on Earth to supply all total energy needs for thousands of years.

The conditions for a fusion reactor to ignite depend on three factors: the density of the plasma n_e, the confinement time τ_e, and the temperature T. In neutral plasmas, the electron density is equal to the ion density. The confinement time measures the rate at which the plasma loses energy, either through heat conduction in the device walls or by radiation (accelerated charged particles

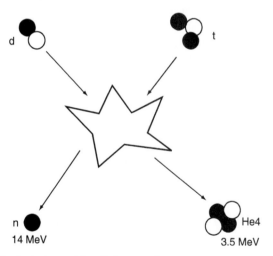

18. **The deuterium–tritium fusion reaction**

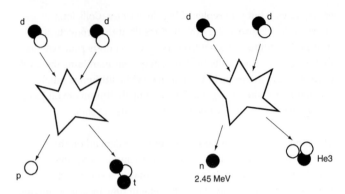

19. The two deuterium–deuterium reactions occur with almost equal probability

emit electromagnetic radiation). For most electromagnetic confinement devices, the fusion plasma must be maintained at a nearly constant temperature. Thus some recycling of energy must be added to the plasma either directly from the fusion products or from the electricity generated to maintain the steady state temperature.

The neutrons cannot be used to maintain the temperature of the plasma because of their lack of electric charge. In the case of

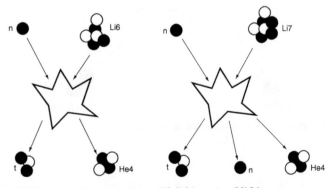

20. Neutron-induced reactions with lithium-6 and lithium-7

the deuterium–tritium reaction, the energy available from the charged particle products is 3.5MeV (Figure 17). Assuming no ions other than fuel ions are present, that is, no impurities, and a 50-50 mixture of deuterium and tritium, the minimum temperature for a sustained fusion reaction is approximately 300,000,000°C, or 20 times higher than that at the centre of the Sun!

The condition for a deuterium–tritium ignition is

$n_e T \tau_e \geq 10^{21} \text{keV s/m}^3$.

No reactor has yet simultaneously achieved sufficiently high values of all three parameters to meet this condition. The current international collaboration ITER (see below) aims to meet this condition on the route to a commercial power plant.

A similar analysis can be applied to inertial confinement systems. In the inertial case, the confinement time must be of the order of the time it takes sound waves (pressure waves) to cross the droplet. This depends on the size of the droplet, the temperature, and the mass of the ions. At a solid deuterium density (0.2g/cm³), ignition would require impossibly powerful lasers. However, if a compressional shockwave can be induced in the droplet such that the density is increased a thousand-fold, then fusion can occur in 1mm-sized droplets. As we have seen, however, it may be possible to achieve ignition at lower densities by the fast ignition technique envisaged for HiPER. In inertial systems, the critical temperature is slightly higher than that in electromagnetic confinement systems.

In magnetic confinement systems, heat must be constantly added to the plasma to maintain its temperature above the ignition point. At start-up, the plasma must be heated to over 100,000,000°C before the fusion reactions start. In the pure deuterium–tritium reaction, insufficient energy is emitted with the charged particles to maintain the plasma temperature. Thus supplementary plasma heating has to be available. The principal plasma heating schemes are: ohmic (resistive heating), heating by neutral beam injection, magnetic compression heating, and radio-frequency heating.

When an electric current flows through an electrical resistor, heat is generated as in the filament electric light bulb or electric fire. The plasma is an effective electrical conductor, and hence a current generated in it will generate ohmic (resistive) heating. However, the resistance of the plasma decreases as the temperature rises and thus the resistive heating becomes less effective. The maximum plasma temperature achieved by ohmic heating alone is around 30,000,000°C and thus additional heat sources are required.

If a beam of rapidly moving neutral atoms is introduced into an ohmically heated plasma, they are rapidly ionized, that is, some electrons are stripped off. The resulting charged particles are trapped in the magnetic field and in their subsequent collisions with the plasma particles they transfer some of their energy, further raising the plasma temperature.

Compression of a gas leads to heating. In the plasma, the magnetic fields drive the ions radially inwards, with the dual effect of heating the plasma by compression and raising the plasma density, which facilitates ignition.

Electromagnetic radiation excites motion in freely moving charged particles, for example in a radio or TV-receiving antenna. Thus high-frequency radiation from an external oscillator such as a klystron with a tuned frequency and polarization will transfer energy to the plasma inducing radio-frequency heating.

There are several technology issues to be dealt with before a power reactor can become a reality. Common to both magnetic and inertial confinement systems are problems associated with large high-energy neutron fluxes and their impact on the structure of the system, the handling of radioactive tritium, and the removal of heat from the reactor core.

The neutron fluxes in the core of a fusion reactor are expected to be more than a hundred times greater than in a fission reactor. These neutron fluxes create two problems. First, they can degrade the material in the core, just as the graphite moderator in AGRs can be degraded; *in extremis*, this could threaten the integrity of the structure. Second, the neutron irradiation can render any materials in the reactor highly radioactive, potentially leading to radioactive waste-management issues similar to those in the decommissioning of fission reactors. After tests at JET with a deuterium–tritium plasma, the largest fusion reactor yet to use this fuel, the vacuum vessel was sufficiently radioactive that remote handling needed to be used for the year following the tests.

However, there is a range of materials of which the reactor components can be manufactured resulting in radioactive waste of much shorter lifetimes (at most a few hundred years) than those from a fission reactor (thousands of years or more for some materials). It is estimated that within 300 years the radioactivity of the waste would be no greater than that from the natural level in coal ash. Design of suitable materials is under way in parallel with the construction of ITER. If the ITER programme is deemed to be successful, then these new materials will hopefully be available for the construction of a demonstration prototype commercial reactor.

Fusion reactors using the deuterium–tritium reaction need a continuous supply of tritium, which it is proposed could be provided by seeding the plasma with lithium; the reactor becomes a tritium generator. Ensuring the safe management of the radioactive tritium is therefore most important. The half-life of tritium, twelve and a half years, is long enough to pose a health risk if the tritium is ingested. The concern is that tritium, being a hydrogen isotope, could contaminate hydrogen-containing food and especially drinking water. This problem could be largely avoided by relying on a deuterium-only plasma. The deuterium–deuterium reaction (Figure 19) produces much less tritium, and the emitted neutron energies are much lower (2.45MeV) than those from the deuterium–tritium reaction (14.1MeV), resulting in less radiation damage. The reaction does not require lithium seeding and hence does not produce additional tritium. It is expected that most of the tritium produced will be burned in deuterium–tritium reactions at the cost of some high-energy neutron emissions. Thus, the tritium handling issues are greatly reduced. The downsides to this proposal are the much greater difficulty in facilitating the deuterium–deuterium fusion reaction and that for ignition the confinement time must be 30 times longer while the power output is nearly 70 times smaller. ITER will study the feasibility of total tritium confinement.

A central problem is how to extract the energy released in the fusion reactions. As 80% of the energy released in the fusion reaction is carried off by neutrons, this limits the ability to use direct energy-conversion techniques. A flow of charged particles represents an electric current from which energy can be extracted by standard magneto-hydrodynamic techniques. Neutrons, having no electric charge, cannot be treated by this method.

The neutrons deposit their energy on the inside wall of the Tokomak. Unless this energy is quickly removed, the temperature could become high enough for the walls of the reactor to melt. On the outside of the reactor vessel are the current carrying coils that generate the magnetic field. To obtain the high fields required, these are cooled by liquid helium or liquid nitrogen cryogenics. The challenge then is to create materials for the wall linings that will absorb the neutron energy and allow the heat to be conducted away to generate power while protecting the cryogenically cooled magnets. The manned space programme faced a not dissimilar problem for the casings of the space shuttle. The protective casing had to be such as to prevent the returning craft being burned up in the atmosphere while protecting the occupants. In both cases, the solution appears to be in the installation of specifically designed ceramic plates.

Unlike fission reactors, there is no possibility of a runaway catastrophe similar to Chernobyl. Any disruption of the plasma conditions would immediately end the fusion cycle. Although the volume of a commercial fusion reactor vessel will be large, 1,000 cubic metres or more, the density of the plasma is small. Typically only a few grammes of fusion fuel will be contained in the vessel at any one time, unlike a fission reactor which is loaded with several years' worth of fuel at a time.

While the growing number of fission reactors has the potential of increasing the risk that their cores may be used to manufacture plutonium for weapons use, there is no direct link between the spread of fusion reactors and nuclear weapons. Although in principle the neutron fluxes could be used to convert uranium-238 into plutonium-239, this would require a major redesign of the fusion reactor.

As a sustainable resource, there is in seawater sufficient deuterium to supply the Earth's current energy demand for 100 billion years and sufficient lithium from which to produce tritium for 60 million years.

Successful production of fusion power holds out the prospect of virtually unlimited supplies of energy for many thousands of years without the emission of harmful carbon gases associated with fossil fuel combustion.

Chapter 8
The need for nuclear power

Energy is the single most important commodity for the preservation of our civilization and the survival of our society. Energy is required to produce the water and food on which we depend. It is required to manufacture fertilizers and pesticides without which food crop yields would be insufficient to support the global population. Energy is required to sow the crops and collect the harvests; it is essential to transport, store, and process our food, and it is needed for cooking. Energy is needed to produce the materials – glass, concrete, steel, other metals, and new substances like plastic – that are essential for every artefact that we manufacture and the construction of all the buildings in which we live and work. We need energy for the heat, light, and power in our homes and in our businesses. Energy allows us to transport people and goods by land, air, and sea. Modern medicine needs energy to produce pharmaceuticals, and to manufacture and operate diagnostic and therapeutic equipment. Energy is the basis of our global communication networks.

There are those who claim all this energy is not required and that we should retreat to the simpler life of yesteryear. While we can certainly use energy more efficiently, a wholesale retreat to a halcyon past is an illusion; the simpler life could support a global population of 1–2 billion people. As we look forward to a global population of 9 billion by 2050, such a retreat would condemn billions to a life of abject poverty and starvation.

The demand for energy is principally driven by two factors, population – the greater the number of people, the more energy they require – and economic activity – the more that they do the more energy is needed to do it. For more than 200 years, since the Industrial Revolution, through wars and economic recessions, global GNP has grown at approximately 2% per annum. Population growth has, until recently, been exponential, although it now appears to be linear. Between 1950 and 2000 the global population more than doubled, from 2.8 billion to 6 billion. Between 2000 and 2050 we are on track for the population to grow to the UN-predicted level of 9 billion. Throughout the 20th century, the population and GDP growth combined to drive the demand for energy to increase at a rate of 4% per annum (see Figure 21). The most conservative estimate is that the demand for energy will see global energy requirements double between 2000 and 2050.

Traditionally, by far the bulk of energy that we use has come from the combustion of fossil fuels. There are concerns that we are rapidly exhausting the finite planetary reserves of fossil fuels, doing great damage to the environment, and that geopolitical circumstances may put our energy supplies at risk. The search is therefore on to find energy sources that will not be rapidly exhausted, will be less damaging to the environment, and over which domestic control is increased.

In Table 4, the percentages of fuels from different primary sources are displayed.

Comparing the global figures with those of the developed OECD nations, we see that heating by burning biomaterials and waste is a feature of less developed countries. The greater use of oil in the OECD countries reflects the greater use of automobiles, and nuclear power is principally a feature of technologically

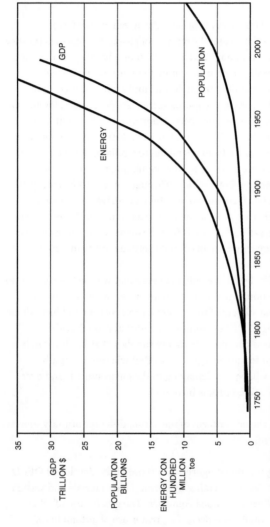

21. The growth in global population, GDP, and energy consumption since 1750. Energy consumption (toe, tonnes of oil equivalent) increases with industrial output leading to GDP growth. Both grow faster than the population, indicating growing global prosperity

	World %	OECD %
COAL	25.3	20.6
OIL	35.0	40.6
GAS	20.7	21.8
HYDRO	2.2	2.0
HEAT	10.0	3.5
OTHER	0.5	0.7
NUCLEAR	6.3	11.0

Table 4. Primary sources of global energy supply (2007). 'Heat' comprises all combustible biomaterial and waste. 'Other' is all renewable sources except biomass

developed nations. The dominance of fossil fuels is obvious from Table 4, with over 80% of our primary energy coming from these sources.

Turning to how we consume energy in Table 5:

	World %	OECD %
COAL	7.1	3.3
OIL	43.0	52.7
GAS	16.6	19.7
HEAT	14.1	3.2
OTHER	3.5	1.4
ELECTRICITY	16.1	19.7

Table 5. Consumption of energy (2007)

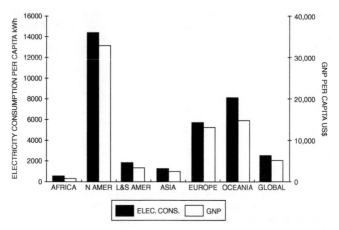

22. A comparison of GNP per capita and electricity consumption; the correlation is obvious

Of note is the greater fraction of energy used in the form of electricity in developed countries. Indeed, electricity consumption per capita is the most accurate measure of material affluence (Figure 22).

Electricity is the growing fuel of choice, in part because a single socket gives the consumer access to heat, light, and power, and in part because it is clean at the point of usage. The demand for electricity is growing at twice the rate of the demand for energy. A comparison of Tables 4 and 5 reveals that the bulk of the coal supply is used to generate electricity, and as such electricity generation is a major contributor to greenhouse gas emissions; the other large contributor to these emissions is oil used in transport. More than two-thirds of all electricity is generated by burning fossil fuels. We use nearly three times more oil than we do electricity (approximately 7% of electricity is produced by diesel generators).

In the 1990s, as concerns about global warming began to grow, the United Nations established a Framework Convention on Climate

Change. The purpose of the Convention was to encourage members to act to mitigate the conditions that were contributing to global climate change. The Convention had no binding powers. In 1997, a group of nations met at Kyoto in Japan, and 35 industrialized countries plus the EU ratified a Protocol legally binding them to meet targets for the reduction of greenhouse gases. The Protocol covered carbon dioxide, methane, nitrous oxide, hydrofluorocarbons, sulphur hexafluoride, and perfluoridocarbons. Detailed rules for implementing the Kyoto Protocol were agreed at Marrakesh in 2001 by 183 parties. Stage one of the programmes was for the Protocol to come into effect in 2005 and to set a target for the reduction of greenhouse gas emissions of 5% in the five years 2008–12. This target was not evenly spread across all signatories. The major obstacle to agreeing targets is the claim by developing countries that, because most of the industrially generated atmospheric carbon dioxide was produced by the OECD nations, that they should carry the highest level of targets for reductions in emissions. Meanwhile, globalization has moved much of the world's manufacturing capacity, a major source of emissions, to the developing world.

The countries that accepted the largest, first phase emission reduction targets to be implemented from 2008 to 2012 were EU 8%, USA 7%, Canada, Hungary, Poland, and Japan 6%, and Croatia 5%.

On a per capita basis, Chinese emissions are less than half those of the USA, but growing rapidly. On the same basis, the greatest emissions come from the Middle East Gulf states, where, for instance, the emissions from Qatar are nearly four times the OECD average of 11 tonnes per person.

Coal and oil each contribute 40% and gas 20% of the carbon dioxide emitted annually from the combustion of hydrocarbons. We see that the dominant sources of carbon dioxide are coal and

	Carbon Dioxide MT/Year	%
CHINA	6,104	21.5
USA	5,752	20.2
EU	3,914	13.8
RUSSIA	1,564	5.5
INDIA	1,510	5.3
JAPAN	1,293	4.6

Table 6. The six largest carbon dioxide emitters produce over 70% of greenhouse gas emissions

oil. Since we use 50% more oil than coal and 25% more coal than gas, we have a clear hierarchy of polluters with coal the dirtiest and gas the cleanest of the hydrocarbons. Pure hydrogen combustion produces only water as waste.

The production of 1GWy of energy from the combustion of coal produces 2.85MT of CO_2. The corresponding figures for oil, gas, and biomass (including waste) are respectively 2.04, 1.63, and 2.72MT of CO_2. Biomass/Waste is claimed to be neutral with regard to carbon emissions since the CO_2 has recently been extracted from the atmosphere during plant growth and hence burning biomass produces no net increase in the greenhouse gas content of the atmosphere. This assertion has recently been challenged.

To discourage the use of fossil fuels, governments around the world are introducing financial instruments to penalize users or to give benefit to those using alternatives. In a search for sources of electricity generation which will be less harmful to the environment, the options are renewables (wind, waves, tides, biomass/waste, and solar), hydroelectricity, and nuclear. In a search for alternatives to oil for transport, the two technologies closest to market are electricity and hydrogen, and the hydrogen is usually produced by electrolysis requiring electricity and heat.

In considering the ability of various energy technologies to reduce the emission of greenhouse gasses, it is increasingly the case that a holistic approach is taken to the total carbon footprint of each technology. It is often argued that, while little in the way of carbon gases is emitted during the operation of a nuclear power station, the large amounts of steel and concrete used in their construction means that there is a significant carbon footprint associated with them as long as the building materials use fossil fuels for their manufacture. Of course, this is true for any technology. Hydroelectric dams can require vast quantities of steel and concrete for their construction, and similarly the need for a thousand giant wind turbines to produce the same amount of electricity as a conventional power station means that they also can have a significant carbon footprint. What is required is a common international standard for calculating carbon footprints consistently applied to all technologies.

A common political aspiration is to reduce carbon emissions by at least 50% by 2050 compared with 2000. At the 2009 UN debate on global warming, the US president committed his country to reducing their carbon emissions by 80% of the 2000 levels by 2050. China acknowledged the need to contribute to global reductions in carbon emissions. However, China is the most rapidly developing industrial power on the planet, and thus, while it may give more attention to the problem, its demand for energy will continue to grow and with it its carbon emissions. If these targets are to be achieved, the first challenge must be greater efficiency in the use of energy and the substitution of alternative sources to the fossil fuels. A widespread goal is to save 20% of energy used by greater efficiency.

	Electricity Generation (TW)						**% Growth**
	2000	2010	2020	2030	2040	2050	
OIL	0.5	0.4	0.4	0.4	0.3	0.3	−0.4
GAS	1.0	1.2	1.4	1.6	1.8	2.0	1.6
COAL	1.2	1.5	1.7	2.1	2.3	2.5	2.1
NUCLEAR	0.4	0.4	0.5	0.6	0.6	0.7	1.3
HYDRO	0.7	0.9	1.2	1.3	1.4	1.5	2.1
WIND	0.1	0.2	0.3	0.5	0.7	0.9	9.1
OTHER REN.	0	0.1	0.1	0.1	0.2	0.3	4.5
TOTAL	4	4.6	5.5	6.5	7.3	8.2	2.0

Table 7. The global installed electricity-generating capacity. Projections 2010–30 are the EIA's 2009 figures; 2040–50 are linear extrapolations

The US EIA (Energy Information Administration) is projecting that total energy demand will grow by 1.9% per year between 2006 and 2030 (consistent with our conservative estimate that energy demand will double between 2000 and 2050). The EIA predicts that the demand for electricity will grow at 2.4% per year between 2006 and 2030, from 18TkWh to 32TkWh. This leads to a projected global electricity requirement of more than 50TkWh by 2050. The growth in electricity-generating capacity (Table 7) takes into account the growing efficiency of the generating plant.

Prior to President Obama's 2009 commitment to reduce carbon emissions, the EIA predicted that coal, which provided 41% of the electricity in 2006, would provide 43% in 2030. They noted in particular the economic benefit of increased coal-fired generation in the USA, China, and India, all of whom have large coal reserves. Such a growth in coal consumption is incompatible with carbon emission reduction targets. On the EIA's projections, coal could be generating 20TkWh by 2050.

Natural-gas-generated electricity was predicted by the EIA to grow at a rate of 2.7% between 2006 and 2030, and projecting forward to 2050 would account for some 11TkWh.

The use of oil to generate electricity (currently around 7%) was not expected to rise due to the projected increasing cost of this fuel.

In the EIA study, renewable energy is the fastest-growing source of electricity. Growing at a rate of 2.9% annually, renewable electricity generation is expected to reach 3.3TkWh annually by 2030, of which it is expected that 1.8TkWh will come from hydroelectricity and 1.1TkWh from wind power. Extrapolating to 2050, we could see renewable energy generation reaching 6–7 TkWh.

As the world's population grows towards 9 billion by 2050, the use of biomass crops to generate electricity will become untenable due to its competition for food supplies. However, the use of waste combustion along with coal will continue and could provide 1–2TkWh by 2050. The development of biotechnology to grow algae from which biofuels to replace oil for transport could be produced, may have the capacity to greatly reduce carbon emissions from this sector. The algae grow in an atmosphere of carbon dioxide and hence have the additional merit of reducing carbon dioxide in the atmosphere. The timescale for such developments is uncertain at this time.

The EIA's projections are incompatible with the desire to reduce carbon emissions. With a reduction in fossil fuel consumption, consistent with emission targets, and no increase in the contribution from nuclear power, the need would be for renewables to generate around 35TkWh by 2050, not the 6–7 TkWh projected by EIA.

Our ability to deal with the increasing demand for electricity while reducing carbon emissions will depend on the rate at which the renewable sources (currently around 20%, the bulk of which is hydroelectricity, of global energy supply) can be made competitive and brought to market, the development of clean-coal technology and the rate of new-build nuclear plant. Failure to press forward on all fronts as rapidly as possible will condemn billions of people to increasing poverty and catastrophic environmental harm.

Hydroelectricity currently provides 16% of our electricity supply. There are a number of large projects that will be completed before 2020. Most notable are the Brazilian and Paraguayan hydroelectric complex on the Rio Parana and the Chinese Three Gorges Project on the Yangtze River. Between them, these two projects will increase global hydroelectricity capacity by two-thirds. Such projects require large rivers and very large catchment areas to form feeder reservoirs.

The largest fully operational hydroelectric plant in the world is in South America (2008). The Parana River forms the boundary between Paraguay and Brazil and joins the Rio Uruguay at Buenos Aires to form the Rio de Plata which empties into the Atlantic. The Taipu Hydroelectric Power Plant has 18 generators that were installed in stages between 1984 and 1991, with a total capacity of 12.6GW. In 2001, work began on the installation of two more generators which raised the capacity to 14GW in 2006, sufficient to supply 90% of Paraguay's electrical needs and more than one-quarter of Brazilian demand. The Taipu Dam created a reservoir covering 1350km^2.

With its dramatic industrial and economic growth, China is constructing the largest engineering project ever attempted, the Three Gorges Dam on the Yangtze River. The project is planned for completion by 2011 and will have a capacity of 22.5GW. As of 2008, the Three Gorges Dam had an installed capacity of 16.9GW and was generating 15.5GW. When work on the dam started,

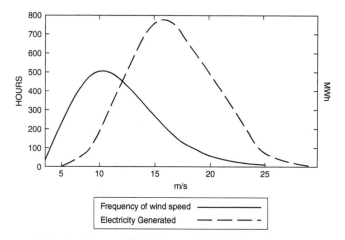

23. The intermittency of the wind means that wind generators do not always generate their maximum capacity. The figure shows the typical frequency, in hours per year, of wind speed and the power generated as a function of wind speed for a wind generator with a 1MW plate capacity. The fraction of actual generation to maximum capacity is called the 'capacity factor'. Average capacity factors lie between 20% and 40%

this would have supplied 10% of China's electricity demand. However, the growth in China's demand has been such that today it would supply only 2.9% of demand and this is likely to be greatly reduced before completion of the project. The dam is creating a reservoir of 632km² but much deeper than that at Taipu. In the process more than two million people will have been displaced.

Both projects have taken over 20 years from conception to full operation. No other comparable projects are currently on the books. It is unlikely that hydroelectricity supplies will more than double between 2000 and 2050. This is consistent with the projections in Table 7.

The most rapidly growing renewable source of electricity generation is wind power, and we have the projection that by 2050 we could have installed around 1TW of wind-power-generating capacity. However, the wind is an intermittent source of electricity. The installed, or 'plate', capacity is not deliverable (Figure 23).

Denmark, which leads the world in the fraction (20%) of the electricity generated by wind power, reported that in 2006 there were 54 days with no wind. They balance their electricity needs using the Norsk grid which allows electricity to be shared by the Scandinavian countries. This illustrates the need for standby capacity to be built when there is a heavy reliance on wind power.

The intermittency of wind power leads to another problem. The grid management has to supply a steady flow of electricity. Intermittency requires a heavy overhead on grid management, and there are serious concerns about the ability of national grids to cope with more than a 20% contribution from wind power.

As for the other renewables, solar and geothermal power, significant electricity generation will be restricted to latitudes 40°S to 40°N and regions of suitable geological structures, respectively. Solar power and geothermal power are expected to increase but will remain a small fraction of the total electricity supply. Large-scale commercial marine power-generating plants are not expected to be available until around 2020, and there are issues of long-term plant maintenance and delivery of power to the mainland still to be addressed. Biomass combustion will be limited because it is less efficient and much more costly than coal and suitable crops will compete for land with the need for food production. The most promising biomass fuel may come in the form of an alternative vehicle fuel produced by growing algae. This has the added benefit that the algae consume atmospheric carbon dioxide.

The world contains coal reserves sufficient for more than 200 years at the present rate of consumption. Coal is inexpensive and widely geographically distributed. On economic grounds, we would expect a steady growth in the use of coal to generate electricity. However, coal is the major source of greenhouse gases, and if targets on carbon emissions are to be achieved, either the amount of coal that we consume must be sharply reduced to less than half the current level or technologies must be developed to deal with the emissions.

Currently, the term 'clean coal' is used to describe coal-fired electricity generation for which the emissions of sulphur dioxide and nitrogen oxides, responsible for acid rain, are eliminated. What is required is a technology that can safely sequester the carbon gas emissions. Currently, approximately 8 billion tonnes of carbon dioxide enter the atmosphere each year due to the burning of coal to produce electricity. This is nearly a million times greater than the waste from nuclear power. Carbon dioxide, being a gas, occupies a much greater volume than solid nuclear waste. Man-made deep geological storage facilities are not a practical solution. Deep geological storage in depleted oil and gas fields is a possibility but much of the coal is burnt far from such sites. Security of underground storage is essential. If the gas leaks into the atmosphere, the whole purpose of the exercise is destroyed. If it is under sea storage and the gas leaks into the ocean, the water will become acidic and the roots of the global food chain could be destroyed.

Experts do not expect fully competitive large-scale wave and tidal facilities before 2020. This means that the bulk of the renewable energy will have to come from wind power.

If, in 2050, half of the renewable electricity generation was to come from wind power, hydroelectric generation was doubled, nuclear capacity was to be phased out, and fossil fuel consumption was halved, then wind power would have to have an installed capacity of 6–7TW. Given that the largest reliable currently available wind turbine has a plate capacity of 3MW and delivers 1MW on average, then at least of 6 million wind turbines would be required, and these would have to be spaced at 200m intervals; the number of turbines required would go round the world 30 times. This unlikely scenario would still leave open the question of standby capacity of around 2TW. To ensure that the standby capacity does not emit greenhouse gasses, this would require some 2,000 nuclear power reactors, but if these could be built it would be totally uneconomic not to use them, as at present, as base load providers with around 90% load factors, because of the high capital costs.

In most industrialized nations, the current electricity supply is via a regional, national, or international grid. The electricity is generated in large (~1GW) power stations. This is a highly efficient means of electricity generation and distribution. If the renewable sources of electricity generation are to become significant, then a major restructuring of the distribution infrastructure will be necessary. While local 'microgeneration' can have significant benefits for small communities, it is not practical for the large-scale needs of big industrial cities in which most of the world's population live.

Supplies of uranium are sufficient for several centuries at present consumption rates. If thorium reactors become a competitive alternative, the supply of nuclear fuel will treble. Weapon-grade stockpiles of plutonium are a rich source of fission fuel for the rest of this century, and if not converted into reactor fuel will remain a potential source of risk for weapon proliferation. With proper decommissioning and waste management, there is minimal risk of environmental damage and no operational contribution to the risk of global climate change.

Following Chernobyl and Three Mile Island, global nuclear engineering capacity sharply declined. Globally, current capacity would allow work to begin on approximately 16 new reactors per year. Assuming an average planning and construction time of 5 years, we could see the appearance of at most 640 new reactors by 2050. However, all of the 439 nuclear reactors currently operating will be in decommissioning by 2050. The maximum number of reactors that could be operating by 2050 is, even allowing for greater efficiency and greater capacity per reactor, insufficient to provide all the carbon-free energy that is predicted to be required.

The targets for the reduction in carbon emissions cannot be met by simply adjusting our means of electricity generation. Of greater importance is reducing emissions from transport fuels. Here the options are limited. Clearly algae-based biofuels are an attractive option. However, it is not clear how long it will be until this technology can be operated on a significantly large industrial scale. Currently available technologies include the further electrification of transport and hydrogen-powered vehicles. However, these would give rise to an even greater demand on electricity.

The bulk of our energy is currently obtained by burning hydrocarbons. The greater the hydrogen to carbon ratio in these materials, the greater the energy output and the smaller the carbon emissions; thus coal (pure carbon) produces less energy and more carbon emissions than the same mass of oil (three hydrogen atoms to each carbon atom), which in turn produces less energy and more carbon emissions than methane gas (four hydrogen atoms for each carbon atom). This suggests that hydrogen is the perfect hydrocarbon fuel; maximum energy output and the only waste product is water. Concerns have been expressed about the contribution atmospheric water vapour makes to global warming. However, we know how to sequester steam by simple condensation.

Hydrogen-fuelled transport is very close to market. There are hydrogen buses on trial in many cities around the world, and Iceland has decided that all transport will be hydrogen-powered by 2050. However, the principal source of hydrogen is water and this requires energy to extract the hydrogen so there is no net gain unless waste energy is used. The greatest single waste of energy occurs in the generation of electricity by burning fossil fuels or in a nuclear reactor. In all these cases, the energy in the fuel is used to generate heat which is then used to drive turbines. The amount of energy in the form of electricity leaving the generator is only one-third of the fuel energy. The rest of the heat is usually dispersed in the atmosphere via cooling towers or flue stacks and, especially in the case of nuclear plants, it is dispersed in the sea, lakes, or rivers. This means that if consumers reduce their electricity consumption by 20%, this saves 6.6% of the fuel energy; meanwhile 66% of the fuel energy is wasted. In some countries, attempts are made to mitigate this waste by using the heat for district heating schemes or local industry. Such plants are known as CHP (combined heat and power) plants and are widely used in Scandinavia and Germany. Globally, they represent less than 10% of electricity-generating capacity. For CHP to work, the consumers of the heat must be close to the generating plant; heat cannot be transported over long distances efficiently. It is unusual to find industrial or domestic consumers close to nuclear plants. The usual means of extracting hydrogen from water is by electrolysis. However, at high temperatures water molecules will split up more easily. The higher temperatures at which Generation III+ and IV reactors operate could mean that their 'waste' heat could provide a basis for hydrogen production.

Electricity cannot be stored in large quantities. If the installed generating capacity is designed to meet peak demand, there will be periods when the full capacity is not required. In most industrial countries, the average demand is only about one-third of peak consumption. As a consequence, non-carbon-emitting electricity generators have spare capacity to produce hydrogen when not on peak load.

If the heat output and spare electricity-generating capacity were to be used by all future nuclear power stations, their efficiency would be doubled.

Without nuclear power, the world will experience catastrophic electricity supply problems by mid-century or will continue to produce the bulk of its electricity by burning fossil fuels with even more serious environmental consequences. However, if nuclear power is to make a significant contribution, investment in construction capacity is urgently needed.

By the end of this century, the renewable technologies and new generating technologies, such as nuclear fusion, may well be sufficiently developed to make nuclear fission redundant. Until that time, thermal nuclear fission power will remain a vital component of global energy supply.

Appendix

The SI (International System) of units provides the joule (J), named for the English scientist James Joule, as the unit of energy and the watt (W), named for the Scottish technologist James Watt, as the unit of power. Power is defined as the rate of generation or consumption of energy.

$1W = 1J$ per second

The exponential numerical shorthand uses powers of ten to express very small or very large numbers and alphabetically the first letters of the Greek and Latin names to designate them:

c (centum) for a hundred, $100 = 10^2$
k (kilo) for a thousand, $1000 = 10^3$
M (mega) for a million, $1,000,000 = 10^6$
G (giga) for a billion, $1,000,000,000 = 10^9$

NOTE: The English billion is a million million, 10^{12}, while the American billion is a thousand million, 10^9. The American numbering has prevailed, but the classic language is used by the international community and hence the designation G (Greek for giant) and not B.

Fractions are designated:

m (milli) for a thousandth, $1/1000 = 10^{-3}$
μ (micro) for a millionth, $1/1,000,000 = 10^{-6}$

Typical electrical ratings: tungsten filament light bulb (100W); domestic electric heater (1kW); large wind turbine (1MW); large gas, coal, or nuclear power station (1GW).

The electrical industry tends to measure energy generation and consumption not in joules but in units of power times the time of usage. Thus the domestic unit of electrical energy is commonly sold in kilowatt-hours.

1kWh = 3600,000J = 3.6MJ

The energy output of a 1GW power station running full out for a year is 1GWy

$1GWy = 3.2 \times 10^{16} J$

In subatomic physics, the unit of energy is the electron-volt. This is the potential energy of an electron in an electric potential of 1 volt (V), named for the Italian scientist Alessandro Volta.

$1eV = 1.6 \times 10^{-19} J$
$1J = 6.2 \times 10^{18} eV$

Using the Einstein equation $E = mc^2$, where c is the velocity of light (3×10^8 m/s), we can define the masses of subatomic particles in energy units. Thus the electron, proton, and neutron masses are:

$m_e = 0.511 MeV = 0.9 \times 10^{-30} kg$
$m_p = 938.25 MeV = 1688 \times 10^{-30} kg$
$m_n = 939.55 MeV = 1691 \times 10^{-30} kg$

The temperature of a gas is a measure of the kinetic energy per particle $k_B T$. Here the Boltzmann constant $k_B = 1.38 \times 10^{-23}$ J/K $= 8.6 \times 10^{-5}$ eV/K. Thus a gas at room temperature (~ 300K or $27\,^{\circ}$C) consists of particles with an average kinetic energy of ~ 0.025eV.

Further reading

J. Polkinghorne, *Quantum Theory: A Very Short Introduction* (Oxford University Press, 2002). An introduction to the quantum theory that describes the behaviour of the subatomic world.

R. Rhodes, *The Making of the Atomic Bomb* (Simon and Schuster, 1986). The most authoritative history of the Manhattan Project.

H. Bethe, *The Road from Los Alamos* (Simon and Schuster, 1991). A personal account of wartime Los Alamos by one of its most distinguished scientists.

R. Jungk, *Brighter than a Thousand Suns* (Harcourt, Brace, 1958). The first popular account of the development of the atomic bomb.

EIA, *Annual Energy Outlook 2009*. The Energy Information Administration's most authoritative collection of data with analysis and forward projections of energy supply and demand for the USA.

EIA, *International Energy Outlook 2009*. Similar in scope to the Administration's national review, the *International Energy Outlook* presents a global review.

IAE, *Key World Energy Statistics 2009*. The key annual study of global energy supply and consumption broken down by fuel, usage, and region.

IAE, *Projected Costs of Generating Electricity - 2005 Update*. An authoritative analysis of the cost of generating electricity.

IAEA/WHO/EC, *Ten Years after Cherrnobyl What Do We Really Know?*, 1996. A joint International Atomic Energy Agency, World Health Organization, and European Commission study of the Chernobyl disaster and its aftermath.

L. Arnold, *Windscale 1957: Anatomy of a Nuclear Accident* (Palgrave Macmillan, 2007). A thoughtful account of the Windscale fire.

NRC, *Fact Sheet on the Three Mile Island Accident*, 1979. The USA's
Nuclear Regulatory Commission's report on the Three Mile Island
incident.

A. Fentiman and H. Saling, *Radioactive Waste Management* (Taylor
and Francis, 2002). A useful introduction to all you need to know
about radioactive waste management.

J. Walker, *Three Mile Island: A Nuclear Crisis in Historical Perspective*
(University of California Press, 2005). Looks back on the Three
Mile Island incident and the impact that it had on the USA and
the world.

Index

NUCLEAR WEAPONS
A Very Short Introduction
Joseph M. Siracusa

In this *Very Short Introduction*, the history and politics of the bomb are explained: from the technology of nuclear weapons, to the revolutionary implications of the H-bomb, and the politics of nuclear deterrence. The issues are set against a backdrop of the changing international landscape, from the early days of development, through the Cold War, to the present-day controversy of George W. Bush's National Missile Defence, and the threat and role of nuclear weapons in the so-called Age of Terror. Joseph M. Siracusa provides a comprehensive, accessible, and at times chilling overview of the most deadly weapon ever invented.

www.oup.com/vsi

SUPERCONDUCTIVITY
A Very Short Introduction
Stephen J. Blundell

Superconductivity is one of the most exciting areas of research in physics today. Outlining the history of its discovery, and the race to understand its many mysterious and counter-intuitive phenomena, this *Very Short Introduction* explains in accessible terms the theories that have been developed, and how they have influenced other areas of science, including the Higgs boson of particle physics and ideas about the early Universe. It is an engaging and informative account of a fascinating scientific detective story, and an intelligible insight into some deep and beautiful ideas of physics.

www.oup.com/vsi

Expand your collection of
VERY SHORT INTRODUCTIONS